Taking Charge of Your Pregnancy

Taking Charge of
Your Pregnancy

THE NEW SCIENCE FOR A SAFE BIRTH
AND A HEALTHY BABY

....

Susan J. Fisher, PhD

with Becky Cabaza

HOUGHTON MIFFLIN HARCOURT

BOSTON NEW YORK 2021

NEITHER THE PUBLISHER NOR THE AUTHOR IS ENGAGED IN RENDERING PROFESSIONAL ADVICE OR SERVICES TO THE INDIVIDUAL READER. THE IDEAS, PROCEDURES, AND SUGGESTIONS CONTAINED IN THIS BOOK ARE NOT INTENDED AS A SUBSTITUTE FOR CONSULTING WITH A HEALTH CARE PROFESSIONAL. ALL MATTERS REGARDING YOUR HEALTH REQUIRE MEDICAL SUPERVISION. NEITHER THE AUTHOR NOR THE PUBLISHER SHALL BE LIABLE OR RESPONSIBLE FOR ANY LOSS OR DAMAGE ALLEGEDLY ARISING DIRECTLY OR INDIRECTLY FROM ANY INFORMATION OR SUGGESTION IN THIS BOOK.

For information about permission to reproduce selections from this book, write to trade.permissions@hmhco.com or to Permissions, Houghton Mifflin Harcourt Publishing Company, 3 Park Avenue, 19th Floor, New York, New York 10016.

hmhbooks.com

Library of Congress Cataloging-in-Publication Data
Names: Fisher, Susan J., author. | Cabaza, Becky.
Title: Taking charge of your pregnancy : the new science for a safe birth and a healthy baby / Susan J. Fisher, PhD with Becky Cabaza.
Description: Boston : Houghton Mifflin Harcourt, 2021. |
Includes bibliographical references and index.
Identifiers: LCCN 2020050867 (print) | LCCN 2020050868 (ebook) |
ISBN 9780544986640 (trade paperback) | ISBN 9780358409076 (hardcover) |
ISBN 9780358611509 | ISBN 9780358611578 | ISBN 9780544986657 (ebook)
Subjects: LCSH: Pregnancy. | Childbirth. | Prenatal care.
Classification: LCC RG551 .F57 2021 (print) | LCC RG551 (ebook) | DDC 618.2—dc23
LC record available at https://lccn.loc.gov/2020050867
LC ebook record available at https://lccn.loc.gov/2020050868

Book design by Laura Shaw Design

Printed in the United States of America
DOC 10 9 8 7 6 5 4 3 2 1

CONTENTS

INTRODUCTION

As a society, we teach our kids about a lot of things, but pregnancy isn't one of them.

I know this because my daughters were in the minority, thoroughly versed in all aspects of reproduction by the time they were teenagers. As they were growing up, my girls loved to talk about what they were learning, new ideas that were springing forth, and the life skills they were acquiring. My husband, who is a physician, and I also discussed our work lives with them. Because my research focuses on human development and pregnancy, these topics were a natural part of our conversations and the girls learned by osmosis.

But how do people — especially pregnant women or those contemplating the possibility — who don't grow up in the home of a pregnancy professional acquire this knowledge? They don't. After raising two girls and meeting hundreds of their friends along the way, I know that most American children are not home-schooled on the topics of pregnancy and birth. This always seemed like an odd omission, since many are aware of the basics of reproduction before they enter kindergarten.

Today most children under age five have some sense of where babies come from. But the openness with which we now discuss the birds and the bees has not spilled over into what happens after conception: how a fertilized egg grows from an embryo into human form and how that fully developed baby is birthed into the world. You would think that learning about what really happens in pregnancy would stand side by side with sex education in our schools, but pregnancy and birth have yet to be granted a prominent position in our curricula. It's time for a change.

There is a growing understanding that some aspects of pregnancy, such as whether a baby will be born early or late, may be heritable. This means that details about how your mother carried and birthed you may be worth knowing as you embark on your own pregnancy. I was highly motivated to educate my daughters about my own pregnancies because my mother was not able to talk to me about any aspect of reproduction. She answered my first fumbling adolescent questions about my changing body by sending me to the dictionary to look up the definition of a single word: menstruation. Is it all clear now? Good.

Similarly, not a word was ever spoken about pregnancy. I was in the dark about the circumstances surrounding my own birth until I was an adult. I was deep into my twenties when my father told me, in passing, that my mother had been in labor for two days before I was born, at least three weeks early. The small-town doctor in attendance could do nothing but stand helplessly by as she struggled. I could tell by the expression on my father's face and the tone of his voice that this had been a harrowing experience. His own birth, I eventually learned, had also been a protracted affair — so difficult that my grandmother, a young woman when she gave birth to him in 1912, avoided ever repeating the experience, so my father remained an only child.

When I became a mother, I was determined to break the pattern of secrecy that seemed to surround pregnancy and birth. My work gave me the knowledge and confidence that are needed to pull these topics out of the dark corners and weave them into everyday conversations. I hope you will share what you learn in this book with friends and loved ones and eventually your own children — who may not yet be born.

THE IMPORTANCE OF TAKING CHARGE
OF YOUR PREGNANCY

What is there to say about the topic of pregnancy that's new? Every chapter you're about to read takes on that key question, and what you learn will influence your decisions in navigating pregnancy and birth. When you're armed with science-based information, you're in a position to take charge of your health care and your baby's well-being before, during, and after pregnancy.

Chapter 1 covers the explosive development of a fertilized egg into a miniature child, which happens during the mere first eight weeks following conception. Understanding the blinding speed at which pregnancy is launched provides an important rationale for preparing your body to take on its critical new role of being baby's first home. There are numerous simple strategies for ensuring it's a nurturing environment, which can up the odds that you'll have a healthy child.

As placental mammals, baby humans develop within the safe confines of their mother's body, nurtured by the placenta. This remarkable temporary organ, the subject of chapter 2, is the baby's lifeline inside the womb. Placental structure and function are explored early in this book because the placenta steers the course of pregnancy, from implantation to birth. During the first month following conception, the placenta begins the process of extending its treelike branches deep into the uterus, a connection that plugs the baby into the mother's organ systems, which it shares for the duration of pregnancy. And after the baby is born, a mother can sustain this new life with her breast milk. Accordingly, we humans belong to the class of creatures known as *Mammalia,* or mammals, terms that derive from the Latin word for "breast." These ancient biological processes, encoded in our genes, are immutable on a human timescale. Yet many other aspects of pregnancy are undergoing change.

Today the how and even the when of baby-making are shifting under our feet. Fifty years ago there were few options other than allowing nature to take its course. Now women and couples have more control than ever in determining when pregnancy will happen. The process involves a sea of choices and decisions, informed by significant cultural shifts in the way we navigate relationships and reproduction.

In 1980, just 18 percent of women who gave birth were unmarried. According to the Centers for Disease Control and Prevention (CDC), the proportion was 40 percent in 2018. The advent of in vitro fertilization in the 1970s, which made pregnancy possible for millions worldwide, also opened the doors to the development and widespread adoption of assisted reproductive technologies such as oocyte cryopreservation, more commonly known as egg freezing. In recent years, we have learned that the age-related decline in female fertility is largely due to a rapid deterioration in the quality of eggs. With the right hormonal treatment and a robust egg donated by a young woman, the childbearing years can be extended far beyond what anyone ever thought was possible. This has inspired more women, most often between the ages of thirty and forty, to hit the pause button on the biological clock by literally placing their eggs on ice, deep in a freezer. Technological advances now allow a woman who wants to delay pregnancy to be her own donor, utilizing her stored eggs for pregnancy when the time is right, rather than being bound to a schedule dictated by her biology. Egg freezing is accelerating the increase in age of mothers who are giving birth for the first time.

When it comes to prenatal care, new technologies abound. Major advances in modern imaging methods now allow assessment of your baby's anatomy — the medical equivalent of counting ten fingers and ten toes — at earlier stages of development. Prenatal genetic testing has also progressed. Until 2011, invasive procedures involving needles or the threading of tiny tubing (amniocentesis and chorionic villus sampling) were the only options. Then, seemingly overnight, these sometimes-risky procedures were replaced by simple blood tests, making the drive to the clinic the most hazardous part of the testing process. In chapter 3, you'll see how this dramatic shift in prenatal testing options is driven by the discovery that a mother's blood contains detectable amounts of her baby's DNA, which can be used for genetic analyses.

What's new about pregnancy doesn't stop with the latest technologies and treatments. There is also updated thinking about the serious effects that chemicals can have on reproductive health, a topic that is covered in chapter 4. Manufacturers and other businesses large and small produce and use approximately eighty thousand different chemical compounds. Everyday life entails countless exposures to them. For example, the thermal paper receipts we receive at the cash register are usually coated in bisphenol A (BPA), a chemical that can disrupt endocrine function — meaning that it can short-circuit the activity of sex hormones, which are essential to reproductive health. Although the problems caused by environmental chemicals can seem insurmountable, there are many easy ways to lower the risk they pose to reproductive health and pregnancy. One is to decline those receipts printed on thermal paper. In the pages of this book, you will learn many similar tips that will benefit you and your whole family.

As pregnancy progresses beyond the first trimester, the chances of developing a complication *increase* — even if things have been normal up to that point. In chapter 5, I'll walk you through what we've discovered about such conditions, which include preeclampsia, preterm labor, and gestational diabetes, and how we're applying those findings to improve pregnancy outcomes. Taking simple steps to ward off these problems can help some women avoid them. Being familiar with the warning signs, which often appear suddenly, prepares you to seek appropriate medical care sooner rather than later, should such a condition arise.

Why are preventing and treating pregnancy complications so important? One of the most surprising findings that has emerged from pregnancy research is that the quality of a baby's life in the womb can spill over into his or her physical and emotional health as an infant, young child, teen, and full-fledged adult. That's right — what you do (or don't do) during pregnancy can impact the health of your child at every stage of life, right through old age. This theory

(known as "fetal programming of adult health and disease," the subject of chapter 6) is both fascinating and empowering. This amazing
piece of science shows that you, as a parent, have the opportunity to
take steps, before and during pregnancy, to positively influence the
health of your child far into the future.

Chapter 7 concludes the book with a detailed description of the
human birth process, which has no real parallels in other animals.
Understanding the biological basis of labor and delivery clarifies
many of the decisions you will want to make before the end of your
pregnancy. Should you consider a home birth? What do experts say
about the different types of pain control? Should you eat the placenta?

I've always believed that the best way to navigate any kind of
uncharted territory is to acquire as much data as possible about the
unexplored landscape spread out before you. In the case of reproduction, a working knowledge of the science of human pregnancy
— which includes much more than egg meets sperm, baby grows in
uterus, woman gives birth — is a strong start, though this information
can be strangely hard to come by. Taught primarily in the first year of
medical school, it's not something you can easily bone up on by taking
an online course or enrolling in a continuing-ed program.

People outside the medical professions are generally left in the
dark. Even my colleagues at the University of California San Francisco
— quite a sophisticated bunch of scientists — regularly come to me
with questions about pregnancy. They know that I used to teach anatomy and that I have spent decades researching human development
and pregnancy, including the many secrets of the placenta. I usually
have answers to their questions, and I hope I'll have answers to yours.

I'm speaking not only as a scientist but also as a mother who has
given birth twice, and my childbearing experiences resemble those
of millions of women. The first time around it was challenging to get
pregnant, and once I conceived, I was hit with a set of new problems
for which I was completely unprepared. This should not have been
the case. Since I was in a high-risk group, I should have been made
aware that I stood a greater chance of encountering these issues. But
that did not happen.

I was born to a mother who was given diethylstilbestrol (DES) to prevent pregnancy complications. Later it was found that the daughters of women who took DES were susceptible to preterm labor, and I did indeed experience this. Although the science of pregnancy was my business, I did not have access to information that could have helped me take charge and avoid the potential difficulties. The second time around I knew better and took a few simple precautions. The result: a healthy pregnancy and a full-term delivery just a few days short of my due date.

This book is meant to provide you with the course in human pregnancy that can be hard to find elsewhere. My aim is to give you the information you need and then turn these facts and statistics into actions that will maximize your chance of having a healthy baby and child. There is a profound connection between my work and your personal experience, between exciting discoveries in research and practical applications that will benefit you and your family. I want to show how medical science can inform and support the primal, all-encompassing experience of giving birth, making it safer for you and your baby. I want you to feel prepared for each step in the process, so that you can take charge of your pregnancy.

1

The First Eight Weeks
Getting Off to the Right Start

Though it takes up the better part of a year, pregnancy is hardly a meandering journey. Human development takes off at mind-bending speed from the moment that sperm and egg collide. The dramatic events that launch the formation of a human body in miniature, from head to toe and everything in between, take place over the short span of eight weeks. This is an especially important period for your baby, and here you'll find out what you can do to make the most of this time.

Fortunately, most babies born in the United States today are generally healthy, as are their mothers. The phrase that describes this typical experience — "uneventful pregnancy" — sounds rather lackluster, considering that it describes a crucially important part of every person's life: the action-packed beginning. The first two months are particularly critical. The fertilized egg, a single cell, grows explosively, rapidly morphing into a baby.

A NEW LIFE, A HEALTHY BEGINNING

There was a time, not all that long ago, when pregnancy was treated almost as an illness, a condition too intimate to discuss frankly and best left to the experts: the (mostly male) doctors who provided scant information, and only on a need-to-know basis. Women were not nearly as well educated about their own health and bodies as they are today, nor did they have many choices about their care. And the honest sharing of some pregnancy experiences that we take for granted now took place only between female relatives or close friends. And even then, it wasn't always viewed as polite conversation.

Fortunately, as the decades passed, societal attitudes changed. Information flowed more freely from doctor to patient and from woman to woman. We loosened up — and that change has been a boon to the health of both mother and baby. Today there is a trove of invaluable, solid science on pregnancy and on best practices for maternal health. This ever-growing body of knowledge has resulted in an array of preventive measures that have already benefited multiple generations.

We know, for example, that good nutrition and physical activity during pregnancy can make you and your baby healthier. Once, pregnant women were encouraged to "eat for two" and stay off their feet. Now we know that inadequate diet, excessive weight gain, and lack of exercise can contribute to complications during pregnancy and delivery. Adding a nutritional supplement (see chapter 2) and taking a daily walk can make a world of difference to you and your baby.

We also know that, unfortunately, numerous toxic substances — environmental chemicals, harmful microbes — can have devastating effects on a baby's development, particularly in the early days and weeks of pregnancy. A mother's use of drugs and alcohol can have similarly detrimental effects. The resulting damage may not be apparent until later in a pregnancy, at birth, in infancy, or even years later.

Whether we're talking about alcohol, a chemical compound, or a dangerous bacterial infection, many such toxic exposures have something in common: they can cross the placenta, the life-sustaining organ that begins to grow in the first week of pregnancy to connect mother and baby. This means that the embryo can be harmed even at the earliest stages of development, before the pregnancy has been detected. (Miscarriage, sadly, can also occur as a result.)

Though it occurred decades ago, the thalidomide tragedy is still among the clearest examples of what can happen when a toxic substance crosses the placenta. Its impact was so immediate. Thalidomide, a drug meant to alleviate morning sickness, was prescribed to thousands of European women starting in the late 1950s. Babies of women who had taken thalidomide were being born with extraordinarily high levels of birth defects, including missing or malformed limbs. This catastrophe ultimately led to tighter screening and regulations on potentially harmful drugs. (The Food and Drug Administration [FDA] blocked its sale in the United States.)

Most pregnant women do follow the cautionary guidelines offered by their health care providers and other reproductive health experts, and as scientists, we've gotten better at identifying toxic substances that can harm a baby in the womb. As a result, we've been able to reduce the occurrence of a variety of pregnancy complications and birth defects. Those of us who closely study maternal health and a baby's development in utero know a lot about what to avoid and why, but there is still much to be learned. And though exposure to "old" toxic substances such as thalidomide is largely a problem of the past, we're still learning about the impact of newer, potentially more complex threats to a developing baby, such as marijuana and the Zika virus.

The research goes on, but for now this much is clear: there are many steps you can take before you become pregnant to up your odds for a healthy outcome. And one of the most compelling reasons for being proactive is the lightning pace of human development that takes place in the first two months of pregnancy, which are known as the embryonic period.

THE FORMATION OF NEW LIFE, WEEK BY WEEK

It's a well-known fact but it never fails to amaze: human pregnancy begins with a single sperm finding its mark — a viable egg. Reproduction is all the more remarkable given that fertility is fleeting, limited to a handful of days each month within a healthy woman's cycle, and made all the more difficult by the problem that most sperm won't survive the journey up the uterine tube, where the egg awaits. For some couples, conception is incredibly simple and may happen without much effort. All it takes is one well-timed meeting between sperm and egg. But for others, the goal is more elusive and fertilization happens in a laboratory dish.

Either way, this lone "founder" cell (or zygote) immediately begins to multiply, and within nine months it will grow from a tiny dot into an instantly recognizable being — a baby girl or boy (or, in some cases, more than one baby). From this one first cell, trillions will eventually form, every one containing genetic material, or DNA, from each parent. Encased in the zygote and its descendant cells, DNA contains the master plan that determines how those cells will function within this newly forming body. If that DNA could somehow be extracted from a cell and stretched out, it would look like a very thin piece of spaghetti just over six feet long. In reality, it does not have a linear structure and instead exists as an elaborate tangle, which somehow unwinds, duplicates, and reassembles every time a cell divides, making a new copy of itself.

Imagine that traffic lights are positioned along the DNA, and they control the movement of the cellular machinery along its contours. If a light is green, the cell "reads" the DNA, and the instructions it contains are carried out by the rest of the cell. During fetal development, these signals change every time a new type of cell is made. A specific pattern of lights directs cells down the street, leading to formation of the heart, for example, while another pattern specifies brain development. During the first two months following fertilization, these routes are heavily trafficked as your baby's body takes shape. (By contrast, for adults, DNA activity slows to the minimal pace that is needed for maintenance mode. To heal a cut on a finger, for instance, we're merely generating new skin cells, not an entire organ system!)

Once the egg is fertilized, the newly formed single cell will divide within the next few hours, and it has to stay on a tight schedule if it's going to make a baby in nine months. The fertilized egg develops into an embryo with two, four, then eight cells and beyond, growing exponentially. Each cell is like a tiny city, with its neighborhoods carrying out specialized functions. During the first few days of the embryo's existence, they all spring into action. Some are sites of energy production. Others are in the import business, bringing in the foodstuffs needed for survival. There are also garbage dumps and recycling centers, which digest spent material that might be toxic to the cell.

This cellular machinery becomes more specialized as development proceeds. Think of New York City and Chicago — they're both cities, but their skylines are radically different. Likewise, cells that are building the heart and the brain are both creating organs, yet those cells have distinctly different functions. This specialization starts early. Research from my group suggests fundamental differences among cells of the embryo only three days after fertilization.

Given the sophistication and specialization of the cells that make up our bodies, it is astonishing how quickly they form and organize. All the big events in human development — including the formation of organs such as the brain, heart, liver, and kidneys — take place within the first eight weeks of pregnancy. The scientific name for the creation of these systems is organogenesis.

Let's look more closely at what happens during the embryonic period, week by week. Keep in mind that organogenesis will be happening well before you show any signs of being pregnant.

Week One
Implantation: When Pregnancy Begins

Development of the embryo starts at fertilization, when DNA from the father's sperm and the mother's egg unite. Only about a hundred or so of the millions of sperm that start the trip actually complete the journey that is required to achieve fertilization — moving up the vagina, through the uterus, and eventually traversing the entire length of the Fallopian tubes (also known as uterine tubes) to penetrate a mature egg. If there is no egg, the sperm can wait around a bit

— in fact, they can live for up to five days in a woman's body. This is why a couple can have sex before ovulation occurs and still conceive.

During ovulation, the egg is released by the ovary. It's swept by fingerlike structures into the cupped ends of the tubes, where fertilization takes place. The egg is encased in a clear outer membrane (called the zona pellucida), a sort of second skin. The first sperm that manages to penetrate this membrane and fuse with the egg causes the zona to instantly harden, which prevents any other sperm from entering. The victor burrows through the egg's inner membrane and enters its interior, where it releases its payload of paternal DNA. This kicks off the formation of the very first cell of what will become a new human body — your baby.

During the next twenty-four hours, this founder cell divides, making a nearly identical copy of itself. This process continues, steadily increasing the number of cells within the membrane. By the fifth day, the embryo resembles a miniature mulberry. In fact, at this stage it's called a morula, the Latin word for "mulberry."

Soon after, cells on the structure's outer surface begin pumping fluid inward, hollowing out a cavity within this tiny round ball. The few cells (about forty out of a hundred) that remain within the ball converge along one side of the interior. The outer cells will become the placenta, a temporary organ that functions like lungs, kidneys, and gastrointestinal tract all rolled into one; those cells that remain

The cells of a five-day-old human embryo, arranged like the seeds of a mulberry, are indistinguishable under a microscope. A protective outer membrane encircles the tiny cluster.

inside, clustered together and pressing against one arc of the developing embryo, will grow into the baby. As it makes its way along the tube and toward the uterus, this microscopic sphere is prevented from attaching anywhere along the glide path because it is enclosed in the zona pellucida, which has a nonstick surface.

Still, occasionally an embryo will latch onto the tube wall, and the result is called an ectopic pregnancy. Because the uterine tube has a diameter of only about one centimeter, with a limited ability to expand, a pregnancy cannot be sustained in it. The attached embryo will rapidly outgrow its narrow confines and may eventually rupture the tube, resulting in a medical emergency. (See page 50.) Though experts aren't precisely sure what triggers an ectopic pregnancy, the risk rises among certain women: those who have had previous ectopic pregnancies, those who have scar tissue or other tubal damage such as that caused by sexually transmitted diseases, and those who smoke.

In most cases, though, the embryo enters the uterus, where it "hatches" out of the tough membrane that originally surrounded the egg, a process that is very much like popping a grape out of its skin. Implantation takes place when the placental cells on the outer surface

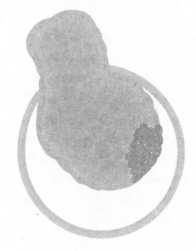

Near the end of the first week of development, the embryo begins to emerge from its protective outer membrane, a process called "hatching." The light gray cells will become the placenta. In just a few hours they will attach the embryo to the uterus, which initiates implantation. The dark gray cells will become the baby.

attach the embryo to the uterus. The optimal position is high up along the wall, where the uterine lining, or endometrium, is thickest and the blood that nurtures development is most plentiful.

This elaborately choreographed dance involves the cells of three individuals — you (your egg and uterus), the father (his sperm), and the newly formed embryo that will become your baby. Pregnancy officially occurs at the moment of implantation, near the end of week one. Implanting on schedule is important; delays are associated with early miscarriage.

Week Two
The Placenta: The First Bond Between You and Your Baby

Those sticky cells on the outside of the embryo are the forerunners of the placenta, a vital though temporary organ that will sustain your pregnancy. The placenta may be the most overlooked yet crucial determinant of your baby's health. It literally forms a lifeline between you and your child, transporting oxygen and nutrients to nourish your growing baby and removing waste products and toxins. A well-functioning placenta keeps you and your baby healthy during pregnancy.

Between the eighth and fourteenth days of embryonic development, the cells that will form your baby continue to multiply, but compared to the rapid growth of the placenta, they are dawdling. At this stage, the embryo is primitive in appearance, resembling a microscopic pancake — a very flat one, as it is only two cell layers thick. But from this simple beginning, something far more complex will emerge.

By the end of the second week of pregnancy, hundreds of placental cells are already beginning to look and act like the complex and sophisticated organ they will become. The placenta will have a major job ahead as both gatekeeper and facilitator, and it is getting a head start — even though you are still unaware that you're pregnant.

Week Three

Body by Design: Laying the Foundation

During the critical days of week three, the "body plan" of your baby is beginning to be established as more and more cells move into position, getting ready to form essential physical features, from skin to internal organs. A three-week-old embryo does not resemble anything most of us would recognize as human, but an incredible amount of detail is beginning to develop.

At this stage, the embryo is shaped like a pear and its cells are arranged in three layers. The outermost layer will go on to form the surface of the skin. Counterintuitively, these same precursor cells also give rise to the brain and spinal cord. The first sign that this process has begun is the appearance of a groove in the middle of the embryo's newly formed "back." This indentation induces the formation of a tube, which will develop into the entire central nervous system. The innermost layer of embryonic cells will become the linings of organs, including those of the digestive system and the lungs. The middle layer forms everything else, such as muscle and bone.

At this early stage, none of these structures look anything like what they'll eventually become. But already some parts of the body — for instance, the head — are growing faster than others, in a process

By day eighteen, the human embryo resembles a flattened pear. The widest part is the future head region and the narrowest, the lower part of the body. The furrow that looks like a "dotted i" in the middle will trigger development of the brain and spinal cord.

called morphogenesis. We humans have big brains, and we need large heads to accommodate them. In fact, the brain keeps growing outside the womb and into early childhood; by age two, your child's brain will be about 80 percent of its adult size. Even when physical growth ends, cognitive development will continue into the late twenties. Already during week three, your baby's future head is the widest part of the embryo, and it will remain disproportionately larger than the rest of the body for the first few years of life. Those big-headed cartoon babies have a basis in scientific fact!

This asymmetrical top-heavy growth causes the embryo to curve. At the same time, the sides of the pear-shaped pancake come together, as if a piece of paper was being folded into a tube. By the end of the third week, the embryo has become a C-shaped cylinder, the perfect home for the tissues and organs that are beginning to develop within. Amazingly the heart has already begun to beat, but it won't be detectable until six weeks.

Week Four

Things Get Busy: Assembling the Human Form

From now on, every week of the embryonic period is jam-packed with landmark events, and week four is particularly significant because a process called segmental organization kicks in. By day twenty-eight, the embryo is no longer amorphous: it is beginning to take on a distinctive shape. Starting toward the head region, small pairs of triangular structures, easily visible as tiny bulges, appear on either side of the midline along the back of the embryo and rapidly increase in number. In a little more than a week, they spread down the entire back. In total, more than forty pairs of these "packets," or segments of differentiating cells, form, eventually becoming your baby's bone, muscle, and deep skin tissue. The timing of their appearance is so precise that scientists who study development determine the age of an embryo by counting the number of these structures.

One way to visualize what these segments will eventually turn

By day twenty-eight, the end of the first month of development, many parts of the baby's body are beginning to take shape. It's easy to distinguish the head from the curvy tail, a remnant of evolution that will soon disappear. Swellings in the head will become the brain, and the large protrusion in the midfront region will form the heart. The chain of tiny bulges on the embryo's back will form the structures of the torso: the deep layers of the skin as well as muscles and bones.

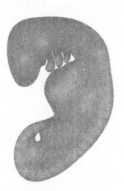

into is to look at someone's bare back. Our thirty-three vertebrae and twelve pairs of ribs are formed from these cell packets, as are the muscles that connect these bones and help them move as a group, along with the overlying skin. Touch a single vertebra, and use your fingers to follow the ribs all the way to the chest. You have now traced out a segment derived from one of these original packets. Although this patterning — all derived from segmental organization — unfolds in other areas of the body during development, it becomes scrambled along the way, making it harder to see the symmetry that is so evident in our torsos.

During week four, the contours of other body parts are also becoming visible. At the top of the embryo, three beadlike swellings will become the brain. Other bulges toward the front of the head region signify the start of the process that forms the lower face and neck. The first bump to appear will turn into the upper and lower portions of the jaw. One of the many remarkable aspects of human development is that some of the events that take place replay our evolutionary history. In the beginning, these jaw bulges look like the gills of fish, from which they evolved.

Other parts of your baby's body that take shape during this busy week include the heart, which develops out of a swelling at the front of the embryo. Two little cup-shaped structures on either side of the head will become the eyes, and the two small indentations below them will turn into the ears.

Weeks Five and Six
Metamorphosis

At this point, a major transformation takes place. The protrusions that form the upper limbs sprout and rapidly elongate into arms. They end with paddles from which fingers are quickly sculpted. Formation of the lower limbs rapidly follows. Internally, meanwhile, organ development is keeping pace. The lung buds branch into the respiratory tree. The long tube that forms the digestive system now includes an enlarged area that will become the stomach, as well as buds that will give rise to the liver and pancreas and fast-growing twisty coils that will form the small and large intestines. These loops lengthen so rapidly that they cannot be contained in the embryo's small abdomen. They protrude and are not reeled back in until the end of the third month.

Another remarkable aspect of the developing gut is that it serves as the highway that the gametes — reproductive cells destined to become either eggs or sperm — will take to reach the ovary or testis. The kidneys go through a mysterious process as well; primitive versions of these organs form and disappear or turn into other structures twice before the real ones begin to grow.

Whether you agree that the developing lentil-sized embryo resembles a tadpole, or you see a seahorse or a bumpy comma instead, at five to six weeks, your baby is changing daily, rapidly attaining a more recognizable form.

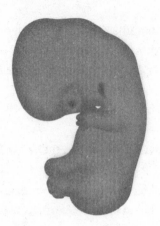

By day forty-eight, the pattern for the baby is nearly complete. The limbs have tiny digits. A primitive eye and, just below it, a slitlike ear are clearly visible. Inside are miniature versions of all the organs. From now on the mandate is growth and maturation.

Weeks Seven and Eight
Your Baby Takes Shape: From Embryo to Fetus

By the end of the eighth week of development, the embryo looks decidedly human. A dazzling transfiguration has taken place, origami of the highest order. A miniature baby is unequivocally appearing.

A name change signifies this pivotal point in the pregnancy. The embryo is now referred to as a fetus. During the next seven months of pregnancy, the structures established during the first eight weeks will grow and mature in preparation for your baby's life outside the uterus.

STEM CELLS: HOPE FOR TREATING DISEASES

Cells that can be coaxed to form any part of the human body are called stem cells. Like the little clump that remains inside the embryo during the first week of development, they are a blank slate. But given the right instructions, they can become one of the highly specialized components of the heart, brain, or pancreas, to name just a few of the cell types they can turn into. The field of regenerative medicine aims to cure serious health problems that are all too common — diabetes, liver and heart failure, as well as optical, muscular, and neurological disorders, including traumatic spinal cord injuries that cause paralysis — by transplanting stem cells to the site of damage. Their chameleon-like properties may eventually enable these cells to morph into replacement parts, curing what would otherwise be a chronic, degenerative, or even fatal medical condition.

PROTECTING YOUR BABY: WHAT YOU CAN DO NOW

For many women, the hormonal shifts of pregnancy trigger nesting behaviors — cleaning, organizing, getting ready. In reality, your baby's first "nest" is your body, the environment where the earliest stages of human development unfold, and there are important steps you should take to prepare yourself, preferably before you become pregnant.

Alcohol: Just One Drink?

Why do doctors always tell women not to drink? And why is it that women in other countries such as France and Italy don't seem to fret over having a glass of wine or a few sips of champagne?

Sometimes the partner of an expectant, teetotaling mom will approach me and urge, "Tell her that having one drink is not going to hurt anything!" My answer is always the same: Don't drink. Alcohol can have harmful effects on a developing embryo, resulting in problems such as premature birth, low birth weight, and miscarriage to name a few. That's because alcohol is classified as a teratogen (tuh-RAH-tuh-juhn): a substance that can cross the placenta, passing from you to your baby, and possibly causing one or more birth defects. In short, it's poison.

Teratogen has Greek roots; *teratos* is Greek for "monster," and combined with *gen*, the root of "generate," *teratogen* essentially means "monster maker." The word is frequently used in categorizing substances that can pose risks to a fetus. Besides alcohol, other types of teratogens include certain drugs, infectious diseases, chemical exposures, and radiation.

The consumption of alcohol in any amount is linked to fetal alcohol syndrome. Its consequences — brain damage that gives rise to learning disabilities, emotional problems, and delayed physical growth — are irreversible. Fetal alcohol syndrome may not be evident until your baby becomes a toddler, when signs of delayed speech and motor development emerge, impediments that can follow your child into adolescence and even adulthood. Birth defects such as problems with the heart, hearing, vision, and more are also associated with a mother's alcohol usage during pregnancy. The American Academy of Pediatrics, the American College of Obstetricians and Gynecologists, the US Public Health Service, and other major organizations that offer science-based prenatal guidelines take the same stance: *During any stage of pregnancy, there is no level of alcohol that can be considered safe.*

But, you may be wondering, what about the women in other countries, where table wine is a given? They seem to drink and produce healthy children. The frustrating answer is that some pregnancies — and we don't know which ones — are highly susceptible to the negative effects of alcohol. And since we don't completely understand the factors that account for this disparity, we don't know how to categorize the risk.

So, if you're still a bit envious that pregnancies in some European countries seem to unfold in a more relaxed way, note that recently, doctors and other prenatal healthcare providers in Italy — where up to 60 percent of pregnant women report consumption of some amount of alcohol — began distributing updated guidelines that reflect a national effort to curb problems such as fetal alcohol syndrome. (France already does so.) *Non bere,* it concludes. Don't drink.

TAKE CHARGE

Stop consuming alcohol now, even if you aren't yet pregnant but are trying. By avoiding alcohol before pregnancy, you are creating a healthier environment for your new baby. Because alcohol crosses the placenta, there is no "safe" level of consumption, whether you are newly pregnant or days from delivery.

Smoking: Its Impact on Your Baby

Most people know that smoking causes cancer, heart disease, and so much more, but its impact on pregnancy and embryonic development, including the placenta, can also be devastating. When health care professionals take medical histories of women who are trying to become pregnant and who are encountering fertility issues, smoking is a huge red flag — and not just for mothers but for fathers as well.

Women who smoke have fewer eggs, and they are of lower quality, meaning that they are less than 100 percent functional. Low-functioning eggs form lower-quality embryos, which are less likely to implant successfully in your uterus. The result can be miscarriage or other pregnancy complications. But even *before* implantation, cigarettes do their damage. Toxins from the chemicals in cigarette smoke end up in the protective layer of secretions that line your reproductive tract and attach to the DNA of cells that reside there. Sperm are forced to travel through a polluted environment as they traverse this terrain on their way to the egg. Once fertilized, the embryo encounters the same chemically contaminated environment on its way to become implanted in your uterus.

Male smokers have lower-quality semen: lower semen volume, reduced sperm counts, and decreased sperm mobility. And if both prospective parents smoke, it's a double whammy. This is why quitting smoking should be a high priority for you and your partner.

Once you become pregnant, smoking continues to pose a threat. Experts estimate that for every cigarette you smoke each day, your risk of miscarriage rises by 1 percent. You are also at increased risk for having a stillborn baby or a baby who dies shortly after birth if you smoke. One reason is that exposure to smoke harms the placenta.

In fact, researchers on my team have found that the more a woman smokes, the greater the damage to the placenta. Much of what we see leads us to believe that in a womb exposed to the constituents of cigarette smoke, the placenta is starved for oxygen. This makes sense since smoking reduces oxygen levels in the blood. We found similar damage to the placentas of women who were passively exposed to cigarette smoke (that is, who did not smoke but lived with someone who did), although the changes are not as great as those in active smokers.

The most well-known complication of smoking during pregnancy is low birth weight. On average, the babies of women who smoke weigh seven ounces less than they would otherwise. While that may not seem dramatic, it is nearly a half a pound, a significant percentage of your newborn's overall weight. Underweight babies can struggle with vital abilities such as feeding and breathing.

The good news is that smoking cessation can have measurable positive effects on pregnancy outcomes. Quitting before conception or soon after reduces the risk of preterm birth.

What About E-Cigarettes?

Because electronic cigarettes or vaping pens produce no smoky odor, they're more acceptable to some people, and their manufacturers like to tout them as less addictive and "cleaner" than traditional tobacco products. But smoke-free doesn't mean safe. The chemicals — including the highly addictive nicotine that goes into the so-called e-juice used to produce the artificial flavors of e-cigarettes or vaping pens — are completely unregulated.

In 2019 alone, over 2,200 e-cigarette users developed vaping-related respiratory illnesses, and 48 individuals died. It's clear that this habit poses serious health risks over and above those of cigarettes, and most of the causes have yet to be discovered. According to the CDC, researchers are still trying to identify the exact components of vaping that caused these illnesses and deaths. Currently, additives such as tetrahydrocannabinol (THC) and vitamin E are highest on the list of suspects.

Vaping is like sipping a mystery beverage that may or may not contain varying levels of poison — a gamble for anyone, but especially risky for someone who is pregnant or who is trying to be. There's no good argument to be made for vaping in place of cigarette smoking.

> **TAKE CHARGE**
>
> If you are a smoker, it's important to quit, especially if you are having problems becoming pregnant. Sperm count and quality are also at risk, so your male partner should stop as well. If you are already pregnant, smoking ups the risk of complications, including miscarriage or low birth weight for your newborn. While passive smoking does not do as much damage as active smoking, it's best to avoid smoky environments. Similarly, vaping poses serious risks.

The Consequences of Cannabis: Toking, Vaping, and Edibles

Because marijuana can help reduce nausea, some consider it useful for dealing with morning sickness. But is it worth it? Based on the research we have so far, and the fact that marijuana's active ingredient, THC — in today's more potent doses — can tweak placental cells, I don't think so.

Cannabis use is way up — more socially acceptable than ever before and legal for medical and recreational use in some states. According to the American College of Obstetricians and Gynecologists, about 2–5 percent of pregnant women report marijuana use during pregnancy. Here's the typical line of reasoning used to support this behavior: marijuana is natural, a plant often organically grown, and therefore is safer than alcohol or other drugs, which may be made using processes or artificial ingredients that are harmful. It's also perceived as less dangerous than smoking tobacco, which has proven links to heart disease and cancer.

Also, cannabis and its offshoots — oils, tinctures, edibles — are rarely sold with detailed warning labels outlining how using them may affect a developing embryo or fetus. (States such as Colorado,

Alaska, and California, where such products are legal, do require such labeling, however.) In contrast, if you pick up a bottle of wine, you'll see a tough-to-ignore message connecting alcohol consumption to an increased chance of birth defects. Warnings like that were added to all beverages containing more than 0.05 percent alcohol in the 1980s. But this difference in labeling should not be taken to mean that marijuana involves less risk — it just means that we don't yet know conclusively how marijuana impacts pregnancy. Substantial time is required to do the rigorous experiments that provide the factual basis for warnings on packaging.

Still, we have plenty of data to suggest that cannabis poses serious risks during pregnancy. It's important to rethink the concept of marijuana as "pure," "natural," and "harmless." Studies have found that using it can lead to stillbirth, low birth weight, and increased rates of admission to neonatal intensive care units. Some studies also link marijuana use with premature birth.

In researching the ways cannabis use influences the development of a healthy baby, my team of scientists found that the placenta has the same receptors that the human brain does for responding to THC — the psychoactive compound in marijuana that gets you high. We are still trying to more fully understand the consequences of marijuana exposure during pregnancy with the hope that the data will assist obstetricians, midwives, and doulas in counseling women about the risks of using marijuana and its synthetic counterparts during pregnancy.

So far, from research done in animal models, we know that THC crosses the placenta to reach the developing fetus. Emerging data from human studies suggest the consequences of maternal cannabis use during pregnancy include alterations in higher-level cognition and psychological well-being. Studies of affected preschoolers report problems with learning, attention, language, motivation, and decision making.

Today's marijuana is undoubtedly more potent than that of the past, and so it may be even more harmful than previously thought. One study estimates that THC levels have gone from 3.4 percent in

1993 to 8.8 percent in 2008 and even as high as 30 percent in some types of hashish, which is more concentrated because it is produced from the plant's sap or resin, as opposed to the buds and leaves.

Even among those who are using cannabis products to cope with morning sickness, the benefit — overcoming a brief period of nausea — is not worth the risk to the long-term health of a child.

> **TAKE CHARGE**
> Because THC affects the placenta and may have a serious impact on development of your growing baby's cognitive functions, you should avoid marijuana and other cannabis products throughout your pregnancy.

Prescription and Over-the-Counter Medications

Whether you are taking a prescription medication or are contemplating an over-the-counter pain reliever or cold remedy, you need an answer to this question: *What will happen to my baby if I put this drug in my body?*

For legal and ethical reasons it's difficult to conduct tests on pregnant women to determine the effects of the many pharmaceuticals available today, so we can't always give a direct yes-or-no answer as to whether a particular medication will hurt your developing child. Fortunately, the FDA requires prescription labeling about pharmaceutical use in pregnancy.

But for those who want more information than a black-box warning, a wide-ranging resource called the Pregnancy Registries, accessible via the FDA website (see Resources), offers answers. Here you'll find data on the effects of drugs approved for use by pregnant women, and you can read through a summary of risks as well as clinical considerations. You can also research whether a drug influences fertility. In addition, a subsection on lactation reviews how various drugs find their way into breast milk and their possible effects on infants.

Another helpful resource, mothertobaby.org, is maintained by the Organization of Teratology Information Specialists. They provide fact sheets about drug use during pregnancy and breastfeeding, and you can sign up for studies that are helping us understand the effects of specific drugs on pregnant women, including many that are used to treat common medical conditions. If you can't find certain information, you can email your questions to experts or even call and talk with a staff member — and it is entirely free of charge.

Your own doctor and pharmacist can provide guidance, of course, but many doctors don't see pregnant patients until they are six weeks into the first trimester, much later than is advisable for considering how a medication may affect your developing baby.

What About Antidepressants?

If you take antidepressants and are pregnant or trying to conceive, should you stop your medication? Some studies point to a small but increased risk of certain types of birth defects among pregnant women who take antidepressant medications. Maternal use of lithium, for example, has been linked to cardiac malformation in infants, though the incidence is low. Other studies have failed to find conclusive evidence tying prescription antidepressants to babies born with birth defects.

It often comes down to a judgment call, one that is best made with the input of trusted clinicians who can provide information and help you make the right choice. If you're mildly depressed, nondrug therapy might be preferable. If you have a long history of depression, however, and have no doubt that stopping your medication will be harmful, the benefits may outweigh the risks.

Today, approximately 5 percent of pregnant women are on some form of antidepressant medication. Given the trends in treatment, that number could well go higher in the years to come, which should motivate researchers to keep studying the topic so that we can provide more specific guidance.

TAKE CHARGE

If you regularly take a specific medication, whether prescription or over-the-counter, be sure to confirm its safety during pregnancy, since medications can cross the placenta and reach the baby. Your health care provider can guide you, but if you are not yet pregnant or are in the early stages and have not seen your clinician, use the online resources listed here and in the Resources section to get some quick answers.

Infection and Illness: Threats to You and Your Baby

The placenta, as you'll discover in the next chapter, is an essential, powerful protector of a developing baby, filtering out toxins as it transports the life-sustaining nutrients required for growth. It is not, however, a magical force field that can keep out every hostile invader. At one time, we believed that the placenta shielded a developing embryo and fetus from infection. But there are some threats it can't stop.

The microbiome, a community of trillions of organisms, resides peacefully within all of us. This includes the uterus and maybe the placenta, a site that is currently the subject of hot debate. (No doubt you've heard of a healthy gut microbiome, for example, consisting of "good" bacteria that live in the digestive tract — helpful organisms that promote normal digestion but that can be destroyed by overuse of antibiotics, among other things.) When the microbiome is in balance, all is well — but when hostile microbes attack and kill off too many beneficial ones, infection occurs. Imagine this battle unfolding in your body during pregnancy, with a weakened placenta and uterus fighting — and failing — to protect the baby developing in your womb.

Scientists group infectious diseases that can cross the placenta and harm the baby under the acronym TORCH. Some are sexually transmitted, but others, such as the parasite that causes toxoplasmosis, are not.

* **T** — Toxoplasmosis, caused by the Toxoplasma *gondii* parasite found in cat feces and undercooked meat, particularly venison, pork, and lamb
* **O** — Other diseases (such as syphilis, chicken pox, parvovirus, mumps, and HIV)
* **R** — Rubella, the German measles virus
* **C** — Cytomegalovirus, or CMV (especially harmful to people with weakened immune systems)
* **H** — Herpes simplex virus

Some TORCH infections (for example, HIV and CMV) are transmitted through bodily fluids such as semen, blood, vaginal secretions, urine, and breast milk. CMV can even be passed from person to person by saliva.

Though these TORCH agents differ in their modes of transmission, they share a common danger: like many teratogens, they often cause the most devastating effects when the microorganism is transmitted from mother to baby early in pregnancy. TORCH agents can set off a range of complications, including miscarriage, birth defects, and health issues that will follow a baby into childhood and beyond.

Take rubella, for example. Babies infected with this virus are born with a variety of conditions, called rubella congenital syndrome, which can impact the ears, eyes, heart, and brain. Later in life, diabetes and behavioral disorders may appear. After the introduction of the rubella vaccine in the late 1960s, occurrences of the congenital syndrome the virus causes plummeted in the United States and Canada. But rubella continues to be a major issue worldwide. (The CDC estimates that each year a hundred thousand children will be born who were infected with rubella in utero.)

The mosquito-borne Zika virus can also cross the placenta, but we don't yet have a vaccine against it. Zika causes microcephaly — a small head and reduced brain size — as well as other birth defects when an infected woman transmits it to her unborn child. Unfortunately, Zika has spread to the continental United States, though the number of cases here is small in comparison to that of South and Central America, Africa, and Asia. Would-be fathers should also avoid mosquitoes because the virus remains in semen for weeks after infection.

COVID-19 is another virus of concern. Early data suggest that it is usually not transmitted from mother to baby. It will be interesting to investigate why this is the case. However, according to the Centers for Disease Control and Prevention, pregnant women may have an increased risk of severe symptoms and pregnancy complications such as preterm birth.

TAKE CHARGE

Viruses and bacteria can harm your developing baby — especially during the first days and weeks of your pregnancy. Fortunately, it is easy to take precautions through common-sense actions.

• Wash your hands carefully, be mindful about food handling and preparation, avoid people who may have contagious illnesses, and talk to your clinician about the necessity of additional steps, such as having a tetanus booster. (For information on vaccines to consider during pregnancy, see the next section in this chapter.)

• If you have a cat, don't change the kitty litter — that's a job for someone who is not pregnant.

• Be aware of foodborne illnesses, especially Salmonella, which is spread through feces, and listeriosis, which can be contracted through eating unpasteurized and raw dairy products, raw or undercooked meats, raw vegetables grown in contaminated soil, and some processed foods that can become contaminated such as soft cheeses and deli cold cuts. Heat kills the bacteria that causes most foodborne illness, so choose pasteurized dairy products, and make sure meats are cooked thoroughly. Wash all produce carefully.

• Avoid traveling to parts of the world where Zika is a threat — the CDC maintains up-to-date preventive information on its website (see Resources), including maps and travel advisories for pregnant women. Always use a mosquito repellent, which is generally safe to do during pregnancy, and talk to your clinician about your travel history or upcoming plans. Also, use a condom during sex if your male partner has been exposed to Zika — the virus can be transmitted through semen.

VACCINATIONS TO SAFEGUARD YOUR PREGNANCY AND YOUR BABY'S HEALTH

Unfortunately, a lot of misinformation has been spreading about the benefits and safety of vaccines — not just for babies and children, but for pregnant women as well. Here are some typical questions about vaccines that have arisen recently:

"Are pregnant women supposed to get the flu vaccine? Or any vaccine? I thought that my doctor wanted me to avoid unnecessary medications and shots during pregnancy."

"Is it the case that vaccines contain thimerosal and other ingredients that could harm me or my baby — or even cause a miscarriage?"

"Is it true that the flu vaccine can give my baby birth defects?"

"With a new flu vaccine coming out every year, how could it possibly be tested to ensure safety?"

"Last year I was sick after I got the flu shot. Now that I'm pregnant, what if getting the flu shot gives me the flu?"

As it turns out, one of the best ways to protect your baby during pregnancy is to stay up-to-date with your own vaccinations. But which ones should you get? Currently the CDC recommends two vaccinations for pregnant women: (1) the inactivated flu vaccine and (2) TDAP.

The Flu Vaccine

Despite the risks that the influenza virus (no matter the strain or branch of the viral family tree) poses to a mother and her unborn baby, only about half the pregnant women in the United States got a flu shot in 2016–17. Those who don't cite several common concerns, most often stemming from a desire to protect their babies.

If you are reluctant to get a flu shot, it's important to talk to your clinician about your concerns. They can answer your questions based on solid information collected from millions of pregnant women who have been vaccinated. For example, we know flu shots do not cause birth defects or miscarriages. It's true, however, that they are not always effective. This is because the manufacturers have to make an educated guess about the dominant strain that will circulate during the coming season. Sometimes they miss the mark, but often they get it right.

Why is getting a flu shot an important part of protecting your baby? For reasons that we don't completely understand, pregnant women who get the flu are often sicker than the general population, with more frequent hospitalizations for serious complications that are bad for mother and baby. This was also true for the coronaviruses that caused severe acute respiratory syndrome (SARS-CoV) and Middle East respiratory syndrome (MERS-CoV). This may also be the case for COVID-19 (SARS-CoV-2), but it will take time to gather the data needed to support this preliminary conclusion. Another advantage of getting a flu shot is that when a mother is vaccinated, the placenta passes on the immunity she develops to her baby, offering critical protection before birth and for weeks after.

There are two types of flu shots. One contains a dead (inactive) virus that can't cause an illness. Known as the "inactivated flu vaccine," this type of immunization is the one recommended for pregnant women. The suggested timing is preconception or as early as possible during flu season. There are, however, some women who should not get this vaccine: those who previously have had a severe allergic reaction to an influenza vaccine or to eggs, which are used in its production.

The other type of flu vaccine is a weakened form of the live virus, which is given in the form of a nasal spray. This vaccine is not recommended for pregnant women.

TDAP

The name of the second recommended vaccination is an acronym for the diseases it protects against: tetanus (lockjaw), diphtheria, and pertussis (whooping cough). Like the inactivated flu vaccine, it contains dead or inactive microorganisms, in this case, bacteria. TDAP can be given at any point during pregnancy, but optimal timing is between twenty-seven and thirty-six weeks of gestation. Women, pregnant or otherwise, who have previously had a serious allergic reaction to the vaccine should not get it. The booster version (Td) is typically given every ten years. It is also considered safe if, during pregnancy, you need additional protection against tetanus, perhaps as the result of a wound, or from diphtheria, a disease that is spread from person to person.

Other Vaccines to Consider

Talk to your health care provider about vaccines for pneumococcal disease, meningitis, and hepatitis A and B. And if you are traveling abroad, it's important to get any vaccines you'll need. The measles-mumps-rubella vaccine (known as MMR), as well as the varicella (chicken pox) vaccine, should not be given during pregnancy, though many experts recommend getting them as part of pregnancy planning, before you conceive.

KNOW YOUR BLOOD TYPE TO PROTECT YOUR BABY

Blood type is designated by a letter (O, A, B, AB) followed by a word (*positive* or *negative*) indicating the presence or absence of a second marker known as rhesus-D (RhD). For example, you may be O-positive, and the father of your child may be B-positive. Your baby is certain to have one or the other blood type. In this instance, his or her health will not be affected by parental blood type because you are both positive for the second marker. A problem arises when there is a mismatch in the second marker — for example, if you are RhD-negative and the father is RhD-positive. If the baby has the father's positive blood type, this could trigger an unwanted maternal immune reaction, similar to what happens when you encounter an infection. Usually this reaction does not affect the mother or first pregnancies.

The placenta keeps fetal and maternal blood separate, but during pregnancy and birth, a small amount of your baby's blood (red blood cells) leaks through microscopic tears in the placenta into your circulation. If you have never before encountered RhD, your immune cells will begin making antibodies that the placenta then carries to your baby, whose RhD positive blood cells they coat — marking them for destruction and causing a severe anemia that can lead to illness and even death in subsequent pregnancies.

Fortunately, it's easy to keep your baby safe. Simple blood work will reveal your RhD status. If you are negative, you will receive a RhoGAM injection, which consists of antibodies that block your immune system from recognizing RhD. This will be done regardless of the father's blood type because the consequences of a mistake are so dire. Typically, in a first pregnancy, RhoGAM is given as an injection near the twenty-eighth week of gestation and within three days following delivery. RhoGAM should also be administered in subsequent pregnancies as there is no way to "cure" RhD incompatibility between a mother and her unborn child. The protection RhoGAM offers lasts for only about three months.

GENETIC SCREENING FOR YOU AND YOUR PARTNER

There's one more item you should consider adding to your to-do list: genetic testing. You may be surprised to learn that experts advise that women undergo genetic testing *before* becoming pregnant.

* *Who should get tested?* You and your partner, since one or both of you could carry a genetic condition that could be passed along to your baby.
* *What's involved?* Generally it's done through a blood test.

These tests, called carrier screening, determine whether a genetic problem is lurking in your DNA. Since we have two copies of genes, we can have one copy of a disease-causing gene but be asymptomatic, because the other copy of the gene is normal and thus protective. However, if the other parent also has one copy of the same disease-causing gene, this can spell trouble. If, at fertilization, the baby inherits two defective genes that cause a particular disease, the child will get the disease.

That's why it's good to determine your carrier status prior to pregnancy. If the test shows you're not a carrier, then nothing more needs to be done. If you turn out to be a carrier, then your partner's DNA can be analyzed to determine whether the disease-causing version of the gene is present. If the answer is yes, there are strategies for preventing a child from inheriting a genetic condition. For example, through in vitro fertilization using your own eggs and your partner's sperm, you can select a specific embryo or embryos that won't develop the disease. The embryos are tested before they are transferred.

The American College of Obstetricians and Gynecologists recommends that anyone contemplating pregnancy or already carrying a child be screened for the gene mutations that cause cystic fibrosis, a disease that makes it difficult to breathe and digest food, and spinal muscular atrophy, a severe degenerative muscle disease that can be fatal before a child reaches two years of age. They also recommend that women be screened for the genetic blood disorders sickle cell anemia and thalassemia.

More targeted screening may be recommended if one or both parents have a family history of genetic disease such as fragile X syndrome, a mutation in a gene on the X chromosome that leads to developmental disabilities. In addition, race and ethnicity can be a factor. For example, people of Ashkenazi Jewish, French Canadian, or Cajun descent are more likely to be carriers of a gene mutation that causes Tay-Sachs disease, a rare and fatal neurodegenerative condition.

You'll find additional information on prenatal genetic testing in chapter 3, as well as a longer discussion of screening and diagnostic tests you may choose to have once you are pregnant. Remember that these are recommendations — not rules. You are in charge of your pregnancy, including whether or not you wish to have carrier screening.

IT'S NEVER TOO EARLY — OR TOO LATE — TO TAKE CHARGE

Make an appointment with your doctor as soon as you decide that you want to start a family. A physical exam could reveal health problems, such as elevated blood pressure, that could be aggravated during pregnancy. This is also the time to evaluate the medications that you and your partner take and get recommendations on their use in the preconception phase and during pregnancy and lactation as well.

If you smoke tobacco or use recreational drugs, including alcohol and marijuana, now is the time to quit, *before* the most vulnerable period of embryonic development begins. If you are already pregnant — beyond the first eight weeks — and are reading this, don't punish yourself if you haven't yet made these changes. Any time is a good time to start making healthier choices during pregnancy.

Here's the bottom line: the odds of having a normal and uneventful pregnancy are in your favor, particularly if you do preconception planning and take special care in those first eight weeks. Not every factor that can impact a pregnancy is under your direct control, but as you can see from this chapter, many are. It is well worth acting upon them.

2

The Placenta

The Lifeline Between You and Your Baby

E ven if you've educated yourself on the physical changes that
accompany pregnancy, as a mother-to-be you'll likely marvel at
your body's dramatic transformation, inside and out, once it's under-
way. You may be wondering how your uterus expands to accommodate a new life, how nutrients reach your baby and sustain its life, why
you have swollen ankles, what causes morning sickness, why you have
sudden obsessional food cravings, or whether your cold symptoms
really are worse this year. Many of these phenomena can be attributed
to the placenta. This long-overlooked, extraordinarily complicated
organ exists for one simple reason: to ensure that you will have a
normal pregnancy and a healthy baby.

Yet doctors and other health care providers rarely spend much
time educating their pregnant patients about the placenta. If your
pregnancy goes smoothly, you likely won't be encouraged to give the
placenta much thought until childbirth, when you'll experience a second set of contractions that mark its delivery. This "afterbirth" is generally disposed of as medical waste, but some new parents want to see
this mysterious blood-engorged organ that kept their child alive for
so many months. Usually the reaction in the delivery room is "That's
gross." Then all the attention is refocused on the brand-new baby.

When I was a graduate student studying human anatomy, one of my professors explained that little research had been done on the placenta. There was scant interest in funding studies on this temporary organ that lives for the mere nine months of pregnancy, as opposed to the mighty heart and the fascinating brain, organs that function for a lifetime. Illustrations of the placenta in anatomy textbooks were minimal and confusing, and the accompanying text was equally baffling. Where did this pancake-shaped organ come from? Was it part of the baby or made by the mother? Did it move with the baby inside the uterine walls, or was it an anchor of sorts? Its functions were summed up as a "life-support system" for the developing baby — so was it some sort of fancy respirator, an aqualung?

I knew I would be unable to learn much about the placenta if I used the standard lab approach of laying the lifeless organ flat on a steel table and teasing it apart to get at its internal architecture. In life, the organ is suspended in a constant stream of maternal blood. To animate the placenta's structure, I needed to re-create this stream.

I found a huge glass dish in the anatomy lab. I filled it with liters of saline, water that has the same salt concentration as blood, because I wanted my experiment to mimic as closely as possible the conditions in the uterus. I swirled a placenta in this pool, and its structure came alive. Intricate branches separated from one another, rippling in the saltwater like trees in the wind. This was the side that had once been deeply anchored to the uterus and interconnected with the mother's blood vessels. When I turned the placenta over, I saw an entirely different surface. The side that faced the baby was covered by a shiny bluish-white membrane that held the remains of the umbilical cord.

Under a powerful microscope, I could see even more detail. The outer surface of the placenta, the branches that originally faced the uterus, was made up of finger-like projections invisible to the naked eye. No other human cells are covered by this dense jungle, which maximizes the placenta's ability to exchange myriad substances with maternal blood. They take up oxygen from the air a mother breathes and nutrients from the food she eats and transport these substances to her baby, who uses them as building blocks for its rapidly developing body. Spent material travels in the opposite direction: back through the placenta to the mother's blood.

To fully appreciate the placenta's central role in steering a healthy pregnancy, it's important to understand how quickly it develops and why this process has a major impact on a baby growing in utero. Along the way, you'll discover how you can optimize the health of this remarkable organ — the only one we humans possess that is generated when it is needed and is dispensed with when its work is done.

A BASIC FACT: NO PLACENTA, NO PREGNANCY

Miscarriage is a deeply sad experience. Because it is so hard to talk about it with others, we may miss the fact that many, many women go through it. Busy health care providers may not think to explain to their patients that miscarriage is often caused by failed placental development.

Regrettably we are far from understanding all the reasons why miscarriages occur, but they happen more often than you may think. Harvard researchers followed a group of healthy women who were trying to become pregnant. Instead of doing a standard test, they used a much more sensitive version that detected pregnancy from the earliest stages. Twenty-two percent of the women had a positive test at implantation but then experienced a spontaneous loss before pregnancy would normally be detected. In total, the miscarriage rate was 31 percent. Ninety-five percent of these women went on to have babies, which suggests that early pregnancy loss is incredibly common and having a miscarriage does not portend future pregnancy problems.

In some cases miscarriage occurs as the body's way of ending a pregnancy that was destined to fail. Many early losses happen because of a chromosomal abnormality or another catastrophic problem. Often, placental failure is the precipitating event. Without a healthy placenta, pregnancy is impossible. Stopping a failing process early on allows a woman to conserve resources and put her energy into a future successful pregnancy and healthy baby.

> **TAKE CHARGE: SLOW DOWN TO UP THE ODDS**
>
> Exercise is important during pregnancy. But if you've been having problems *staying* pregnant and you regularly participate in rigorous sports or other physical activities that involve considerable jumping or jolting, those movements may be connected to failure of implantation and placental development.
>
> Years ago, I knew a biologist who was also a highly accomplished equestrian. She rode and jumped her horses nearly every day. She confided to me that she had miscarried eleven times. On a hunch, her obstetrician suggested she take a break from the horses to see if that might help her carry a pregnancy to term. Doubtful but desperate, she did so — and a year later she had a healthy baby. It's impossible to know exactly what happened in her situation. She had been able to get pregnant on many occasions. Each time an embryo had implanted, but something was interfering with its ability to stay put, and it's not farfetched to infer that the jarring movements of riding were short-circuiting placental attachment to the uterus.
>
> If you suspect a similar problem, talk to your doctor about dialing back strenuous physical activities that require very intense movement.

HOW THE PLACENTA DEVELOPS

After the egg and sperm unite and before the resulting embryo implants in the uterine wall, the development of the placenta takes off at an explosive pace, growing much faster than the embryo. This is because the placenta is like a support team for the baby; it needs to get out in front to provide everything that's needed for normal embryonic growth. The basic structure of a fully functioning placenta is established within three weeks of fertilization.

Will you feel that moment when the embryo implants and the placenta begins to develop? When implantation occurs, about one in five women describe mild cramping, which makes sense, given that the embryo is plowing into the uterine wall and aggressively burrowing beneath its surface to ensure its survival and a successful pregnancy. Even if you have a twinge at implantation, it's likely you won't feel anything when the placenta begins to rapidly establish itself. That's quite remarkable, given what is occurring inside your body.

Every organ has a population of specialized cells, which carry out unique functions. For example, neurons are the specialized cells of the brain; they power our thoughts and propel our actions. The specialized cells of the placenta are called trophoblasts, a word whose Greek origins point to what these specialized cells do. *Trefo* means "to feed and nourish"; *blastos* means "to shoot out or germinate." Trophoblasts transport nutrients from mother to baby, and in the early weeks of pregnancy, they spring to life in the way a seed grows into a young plant. But before they become the gatekeepers for everything that passes to and from the developing baby, they have other important tasks to complete. Imagine trophoblasts starting out as a microscopic work crew, with a long to-do list — the first job being to jump-start construction of the placenta.

Week One

Within five to six days after fertilization, trophoblasts appear, forming the outer surface of the embryo. This tiny ball of cells is making its way down the Fallopian tube. With implantation as its goal, the embryo enters the uterus and the trophoblasts target a spot high up on the uterine wall for attachment. Quickly these placental cells create a tiny doorway that opens into the wall of the uterus. Within hours, the embryo enters this door, and trophoblasts bury it below the surface in the safe confines of the uterine lining. Hidden from view, the tiny embryo is impossible to find in the comparatively vast expanse of the uterus. This is one reason why there are so few images of implantation-stage human embryos that have developed in a woman's body.

Week Two

During the second week after fertilization, the trophoblasts rapidly make copies of themselves. Simultaneously, the cells organize into fingerlike projections that begin burrowing deeper into the uterus — these structures, called chorionic villi, are the placenta's building blocks. Between the second and third weeks, the chorionic villi start to look more like the branchlike forms that they will become. Support cells migrate into the cores, where their internal structure starts to take shape.

Week Three

Blood vessels begin to form within the fledgling branches of what is known as the "placental tree." Eventually they connect with the veins of the umbilical cord, which carry nutrient-rich blood from the placenta to the growing baby. Once depleted of nutrients and oxygen, umbilical arteries carry the spent blood back to the placenta, which turns the waste material it carries over to the mother. At this early stage, the placenta does not have a reliable supply of maternal blood. A sluggish circulation develops as the trophoblast cells demolish tiny blood vessels at the surface of the uterus. But this small amount of blood cannot fully sustain the rapidly growing placenta. Instead, the chorionic villi are fueled by a complex mixture of nutrients, sometimes referred to as "uterine milk," that are secreted by uterine cells.

Beyond the possibility that you may feel cramping at implantation, is there ever any evidence of these early stages of placental

During early pregnancy the rapidly growing placenta resembles a young sapling with branches called chorionic villi. Blood vessels (light-colored arteries and darker-colored veins) run through its interior, carrying countless substances to and from the baby. With continued development, its miniature limbs will increase in diameter and branch many times over. When fully formed, the placenta's structure has the complexity of a mature tree with numerous main and side branches.

development? For some women, the answer is yes. You may have a vaginal discharge. It will be pinkish in color if it's a small amount of diluted fresh blood or brownish if the bleeding happened earlier and clotting ensued. This discharge might be confused with the start of menstruation, but it's actually a sign that the placenta is doing its job of invading the uterus, breaking blood vessels in the process and causing minor bleeding. If the flow turns into bright-red blood, however, it could signal the start of a period or an impending miscarriage.

At three weeks, the placenta, with its trophoblast construction crew, is beginning to remodel your uterus into a perfect home for your baby, though it's possible you still may not realize that you're pregnant.

THE NEXT PHASE: GETTING YOUR BLOOD PUMPING

Those tiny tree branches — the chorionic villi — are now beginning to extend deeply into the uterus, firmly anchoring the placenta to its wall. As they spread out from their original attachment site, they create large openings in the uterine arteries they encounter. Imagine poking a hole in a high-pressure hose; water will spray from that puncture. Similarly, maternal blood sprays from the breached uterine arteries onto the placenta, which sets the stage for fetal growth.

The placenta will capture the nutrients and oxygen in this blood and transport them to the baby, so the amount of blood the arteries supply really matters. You'll be advised to drink plenty of fluids to help support this process. Although the embryo develops in important ways during the first trimester, its growth does not ramp up until maternal blood is rerouted to the placenta, usually at the juncture between the first and second trimesters. At forty weeks, when your baby is full term and ready to be born, about 120 uterine arteries are delivering over three cups of blood to the placenta every minute! This is all thanks to the industrious trophoblasts, which renovated the blood vessels in your uterus as they constructed your baby's temporary home.

In parallel, your blood volume increases over the course of pregnancy by 40 to 50 percent, which helps your body support two lives. This expansion is thought to be driven, in part, by placental hormones. They also promote the retention of sodium, causing the symptoms that many pregnant women are familiar with, such as swollen ankles. Ironically, these same hormonal changes can bring on a classic sign of menopause: hot flashes. Suddenly you are flushed and feeling uncomfortably warm.

Once the placenta is firmly anchored, with a reliable source of maternal blood, there is less chance for things to go wrong. The fact that miscarriage rates fall after the first trimester is partly explained by this development. But it's also why brisk bleeding during pregnancy — in essence a "leak" in a complex system established by the placenta — is an emergency situation, requiring immediate medical attention.

TAKE CHARGE: HELP THINGS ALONG WITH HYDRATION

When you become pregnant, you'll be advised to consume about ten cups of some kind of fluid every day, including water, tea, and coffee (totaling less than two hundred milligrams of caffeine), fruit juice, and milk. (These are the latest recommendations from the American College of Obstetricians and Gynecologists.)

The reasons are many. One is to keep your kidneys, which cleanse your blood, in good working order. On top of their normal duties, they are filtering waste products from your baby that the placenta delivers to your blood. Another is to help your digestion along, which can slow down during pregnancy due to hormonal changes. Also, your body needs fluids to keep up with the required expansion in blood volume. Step up your consumption of water if you are exercising, live in a hot climate, or are having morning sickness — all of which can cause dehydration. In general, staying well hydrated is a good way to keep you and your baby healthy.

THE PLACENTA TAKES HOLD: PUTTING DOWN ROOTS FOR A HEALTHY PREGNANCY

In the same way that newly planted crops need nutrient-rich soil to sustain vigorous growth, the placenta, like a seed, needs healthy ground in which to grow — and that environment is the uterus. From the beginning, the uterine walls form a safe haven in which the placenta thrives.

This protective environment is the work of female hormones. A normal menstrual cycle is a sort of dress rehearsal for pregnancy. Hormones that regulate the cycle cause the uterine lining, called the endometrium, to plump up, filling with nutrients and other substances in anticipation of receiving a burrowing embryo. When that doesn't happen, menstruation ensues and the uterine lining sloughs away.

But an embryo has the power to stop menses. This preserves the prepared uterine lining, which is pliant enough to allow invasion of the embryo and its expansion as it grows. At the same time, it's firm enough to prevent collapse of the fragile embryo and miscarriage. The endometrium is thickest at the top of the uterus, where it is richly supplied with blood vessels. Therefore, this is the best location for a placenta and the baby it will support. For most women, this higher ground is where the placenta will set up shop.

Finding the Right Address

How is the tiny embryo guided to just the right location? Because uterine attachment happens deep inside a woman's body, inaccessible and invisible to observation and study, how an embryo knows where to go has been one of those seemingly unsolvable scientific mysteries. But sometimes in research, clues come from unexpected sources. My colleague at the University of California San Francisco figured out how white blood cells make a beeline for the site of an infection. After learning about his work, I thought there might be parallels with embryo implantation, which requires trophoblasts to head for the right spot in the uterus.

When an infection occurs, the body's immune system telegraphs the exact "address" of the crisis to white blood cells via complex molecular signals. By decoding these messages, they know exactly where to tumble out of the fast-moving bloodstream. The specialized cells come to a screeching halt right where they are needed and do battle in the affected area. But if these powerful weapons are misdirected, the result can be catastrophic — without the correct address, these fighter cells could easily show up at the wrong location and destroy healthy, normal tissues. (This type of misguided attack happens in autoimmune diseases such as rheumatoid arthritis, as infection-fighting cells attack the joints.)

Imagine a vast neighborhood where the houses all look alike, the streets are nameless, and none of the homes are numbered. It would be impossible to deliver a package to the right address. But for a few days a month, one house displays its full address in neon lights and rolls out the red carpet for a very special delivery. That's essentially what happens during implantation — the cells at the optimal uterine site for implantation provide the correct molecular address to trophoblasts. The same signals that lead white blood cells to a site of infection are used by placental cells to deliver the embryo to the implantation site. But this invitation is extended only during that narrow window in the menstrual cycle when the uterus is receptive to an embryo.

Making this "landing" even more challenging is the fact that the uterus normally contracts several times a minute, painless and virtually unnoticeable compared to menstrual cramping and labor contractions. Imagine the miniscule embryo free-floating through the relatively huge expanse of a constantly moving uterus — it's like trying to land an airplane on an aircraft carrier during a hurricane!

Invasion of the Trophoblasts!

If finding the right address is the first step in implantation, then the next phase is like pulling into the driveway of a new home and unloading the moving van. Once the embryo attaches to the uterus, the trophoblasts begin working their way through its lining. This process is called invasion because, under a microscope, these placental cells look a lot like an invading army, forming a phalanx at the surface of the uterus and spreading out as they move deeper.

How do they accomplish this feat? As scientists, we want to figure out the mechanisms because the answers have practical implications. Infertility, early pregnancy loss, and certain complications (for example, preeclampsia, the sudden onset of high blood pressure; see chapter 5) are associated with inadequate invasion. To study how trophoblasts attach and then burrow into the uterus, I transferred small branches of the placental tree to little dishes that I had coated with a gelatin-like material that simulates the cells' environment in the uterus. I then placed them in an incubator, a specialized cabinet that mimics conditions within our bodies — a humid 98.6° F, with the right balance of carbon dioxide and oxygen.

Twenty-four hours later, the results were crystal clear. The trophoblasts that migrated out from the branches had turned the "lab Jell-O" into Swiss cheese! I had created a model for studying how the

Placental cells in a dish coated with "lab Jell-O"

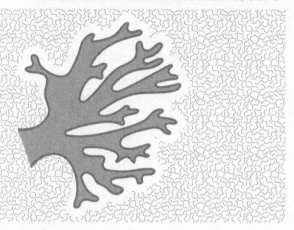

When small pieces of the placental tree are grown on "lab Jell-O" in a dish, the trophoblast cells they contain attach to the surface and "invade" — mimicking their behavior within the uterine wall. They quickly clear wide swaths, visible here as a clear rim around the branches, creating a path forward.

placental cells attached to and invaded the uterus. Since this original discovery, we have used this experimental system to figure out many components of the cells' invasion machinery.

We hope someday to use our understanding of this process to boost trophoblast invasion in cases where a nudge in the right direction could set pregnancy on a more successful course. Additional benefits could include reduced infertility and a lowered risk of miscarriage or complications during pregnancy. And the flip side is equally exciting. It may be possible to develop a new nonhormonal form of birth control based on stopping trophoblast invasion. Many women who have problems with current hormonal methods of birth control have been hoping for such a breakthrough for decades.

OUR LAB JELL-O AND YOUR PREGNANCY

Research using the gelatin model we created in our lab eventually produced data that women and their families can use to make everyday decisions during pregnancy. For instance, we found that when components of cigarette smoke are added to our lab dishes, the placental cells no longer stick as well to the lab Jell-O, and many fewer invade. It's another indication of how active or passive cigarette smoke exposure impairs pregnancy and why avoiding it is so important. Recently we also showed that compounds commonly used as flame retardants change the sticking and invading process. These kinds of data can be used to regulate environmental chemicals with potentially harmful effects on reproduction. (For more on the impact of environmental chemicals on pregnancy, and how you can minimize your risk, see chapter 4.)

The Placenta Grows Up

During much of the first trimester, the branchlike chorionic villi entirely surround the embryo. This is only a temporary arrangement. As your baby develops, it becomes too large to be contained within the narrow confines of the uterine wall and eventually grows into the expandable uterine cavity, a process that uproots many of the chorionic villi from their uterine attachments. These nascent branches degenerate over most of the originally spherical surface of the placenta. As a result, the part of the placenta that remains attached to your uterus is disk-shaped. (In Latin, *placenta* means "flat cake.") In addition to these structural changes, its functional capacity increases. From the second to the third trimester, as growth accelerates, the placenta needs to work harder, matching step-for-step the baby's ever-increasing need for oxygen and nutrients.

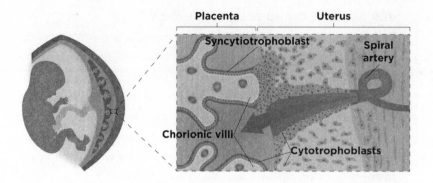

(*left*) As the baby grows, it balloons into the uterine cavity. The portion of the placenta that remains attached to the uterine wall becomes a mature, disc-shaped organ composed of treelike chorionic villi. (*right*) This zoomed-in view shows that cytotrophoblasts (the placenta's resident stem-like cells) either fuse into a large sheet that forms the villus surface (syncytiotropho-blast) or leave the placenta and invade the uterus. They occupy the uterine lining that is adjacent to the placenta, reaching as far as the inner portion of the muscular wall. The placental cells remodel the uterine blood vessels, a process that is particularly vigorous when aimed at the arteries that spiral through the uterus. Cytotrophoblasts replace the mother's cells that originally lined these vessels and significantly enlarge them. This allows the arteries to supply the large amounts of blood required for growth of the placenta and baby.

Know the Location of Your Placenta

While the placenta normally grows near the top of the uterus to take advantage of the plentiful blood supply there, it's not uncommon for an initial ultrasound to show that the attachment site is lower down. In this case, what should you do? While you can't change the position of the placenta, having information about its placement is valuable — one of the many benefits of prenatal care. Having this knowledge can change your approach to pregnancy and birth.

Often the situation rights itself — the placenta's position can change on its own. As it grows and the uterus stretches, the placenta may inch closer to the optimal position, "north" of the baby. But for reasons that are not always entirely clear, the placenta may implant and grow in a less advantageous part of the uterus, at the bottom

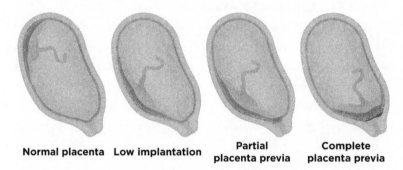

Normal placenta Low implantation   Partial   Complete
placenta previa   placenta previa

Normally the placenta attaches near the top of the uterus. This arrangement enables the baby, which lies closer to the entrance of the birth canal,
to be born first, with the placenta following after. But other orientations are
possible. Sometimes the placenta grows nearer the bottom of the uterus
(low implantation). It may partially or totally occlude the birth canal (partial
or complete previa). In these cases, during labor the placenta detaches and
is expelled first, leaving the baby stranded in the uterus without its oxygen
supply.

rather than the top. Here the uterine lining is thinner, and the trophoblast cells may invade deeper than they are meant to go, particularly into the blood vessels that course through the area. Sometimes
the placenta implants in such a way that it lies partially or totally
across the entrance to the birth canal, a condition called placenta
previa. If this happens, problems can develop. When the placenta is
not properly anchored, it can tear away from the uterus and cause
bleeding. If you have this condition, which is usually discovered
during routine prenatal care, you may be advised to limit your physical activity or be placed on bed rest to reduce the gravitational pull
on the placenta.

The biggest problem with placenta previa occurs at birth. Since it
lies "south" of the baby, the wrong orientation for the placenta, labor
causes it to be expelled first. As a result, the unborn child is stranded
in the uterus without its oxygen supply and its lifeline to the mother.
A discovery of placenta previa is an indication that a cesarean section
may be required. It will allow the baby to be removed prior to the
placenta.

PLACENTA SHAPE: WHY AVERAGE IS BEST

A placenta that is too small or oddly shaped may not be able to support normal development of a baby, and some researchers think that unusual placental shape or size may predict health problems that will emerge when the child grows into adulthood. But it's important not to read too much into the size and shape of a placenta. One commercial laboratory, for instance, claims that basic elements of placental structure can be used to predict a newborn baby's risk of autism, and offers a test to do just that. However, it's nearly impossible to find evidence of abnormal placental structure in the birth records of autistic children, so it's hard to see what real information can be gained from such a test.

The End of the Line

At the conclusion of pregnancy, the placenta's dimensions are impressive. On average, it has a diameter of nine inches and is one inch thick, weighing in at just over a pound. Taking into account all of its treelike branches, it has a surface area of twelve square feet and covers about 20 percent of the uterine surface.

After supporting your baby through pregnancy, labor, and delivery, until he or she is breathing independently, the placenta's job is now over. A few minutes after birth, it separates from the uterus and passes through the birth canal as the mother's uterine arteries constrict to stop the flow of blood. Typically it meets an inglorious end, in a red biosafety disposal bag, after sustaining a baby for months, in myriad complex ways.

WHY A HEALTHY UTERUS MATTERS

A healthy placenta depends largely on a healthy uterus. This organ, normally the size of a small pear, gradually expands until it's eventually able to accommodate a full-grown baby and placenta. And it's not just the mechanical abilities that are so impressive. At the same time its lining becomes transformed into the specialized cells that support implantation and placental development. Given the uterus' varied and pivotal roles in pregnancy, it makes sense to address conditions that impact its function *before* you become pregnant.

Fibroids, STDs, and Other Conditions

A fibroid is a benign growth of uterine muscle that can potentially cause problems during pregnancy. These lumpy balls can be painful and may cause bleeding, or they can go undetected if they don't cause symptoms. Like boulders that make it hard to till and plant a field, fibroids can pose difficulties for an embryo trying to attach itself to the uterine wall. If an embryo does implant, fibroids may interfere with the trophoblasts' reconstruction of the uterus and the growth of the placenta.

Also, sexually transmitted diseases (STDs) and other infections can damage the lining of the reproductive tract, including the uterus and Fallopian tubes, leaving scar tissue that can interfere with implantation and with fertility itself. The incidence of some STDs — chlamydia, gonorrhea, and syphilis — is on the rise. Of the three, chlamydia, which can cause pelvic inflammatory disease in women and is largely asymptomatic in men, is the most common. According to the CDC, there were 1.7 million cases reported in 2017, but experts believe the real number is much higher. Fortunately, it is simple to test for chlamydia, and it can be treated with a single course of antibiotics.

Human papillomavirus (HPV) is the most common sexually transmitted disease. Worldwide, the risk of at least one infection is about 50 percent among men and women. HPV causes problems such as anogenital warts and cervical cancer. Fortunately, vaccines against this virus, such as Cervarix and Gardasil, are decreasing rates of infection and associated diseases.

Endometriosis, abnormal growth of the uterine lining, can also cause fertility problems. Normally, endometrial cells die and are shed during menstruation every month. But in some women, they continue to grow, disrupting normal uterine functions. Sometimes these rogue cells migrate out of the uterus and clog the Fallopian tubes. Or they escape the reproductive tract altogether and grow where they were never meant to be — for example, on parts of the intestinal tract. Somehow, though operating from a distance, they impair the functions of the normal endometrial cells that stay put, which in turn interferes with implantation and placental growth.

Endometriosis can be a painful condition, causing a pattern of irregular periods, pelvic pain, and heavy menstrual bleeding. However, like fibroids or STDs, endometriosis may in some cases have no obvious symptoms. Some women may not even know they have it until they visit a doctor because of difficulty becoming pregnant. Endometriosis can be treated with hormone therapy, but if it is impairing fertility, surgery may be required to remove the islands of cells that are growing abnormally.

Ectopic Pregnancy: One Consequence of an Unhealthy Reproductive Tract

When the embryo hatches and the trophoblasts on its outer surface are set free, they can stick anywhere. Sometimes they get stalled in the Fallopian tube and burrow into its thin walls, meaning that the embryo implants *outside* the uterus. This is called an ectopic pregnancy, from the Latin prefix *ecto-*, meaning "outside." Scar tissue from infections such as STDs or from endometriotic lesions can set up a roadblock in the tube, which prevents the embryo from reaching the uterus.

Ectopic pregnancies, also called tubal pregnancies, are not viable because the narrow Fallopian tube cannot accommodate the rapid growth of the placenta, much less that of a baby. An ectopic pregnancy can be challenging to diagnose because its associated pain can be misconstrued as a gastrointestinal problem, such as appendicitis. A definitive diagnosis usually requires imaging. When detected early, ectopic pregnancy can usually be terminated with minor surgery or with a drug that induces pregnancy loss. The goal of both approaches is to do as little damage as possible to the Fallopian tube. If an ectopic pregnancy is detected later, or if there is a concern about excessive bleeding, then more invasive surgery may be necessary.

Occasionally, the embryo does not enter the uterine tube but instead goes the opposite way — into the abdominal cavity. An abdominal pregnancy is a very rare event and extremely dangerous, due to the risk of bleeding. For this reason, they are terminated.

TAKE CHARGE: GET SCREENED BEFORE YOU GET PREGNANT

Before you try to conceive, it's wise to rule out common issues that can interfere with uterine health and fertility. Fibroids, STDs, or even endometriosis won't necessarily stop you from becoming pregnant, but they could make getting there and having a healthy baby a lot trickier. The condition of the endometrium can determine where the placenta positions itself for growth. The best spot is at the top, but if that area is marred with impenetrable scar tissue, the placenta may look elsewhere for an easier option and implant in a less optimal position. And even if fetal development progresses normally, poor positioning of the placenta could mean complications in later weeks and months.

How the Uterus Expands

Before pregnancy, the uterus sits near the floor of the pelvis and measures about three inches (7.5 centimeters) in length. By the end of pregnancy, it is many times that size, with the top reaching the bottom of the breastbone. If necessary, the uterus can grow even larger to accommodate multiple babies and their placentas.

At about twenty weeks (sometimes sooner), your doctor or other health care provider will start to track the size of your belly, measuring (in centimeters) from the top of the pubic bone (where the hips join in the middle) to the top of the uterus, a distance called fundal height. Tracking uterine expansion provides a window into how the baby is growing. At that stage, your baby will be about the size of a large tomato, and at thirty-eight weeks, the size of a small watermelon. The fundal height measurement should follow a roughly consistent growth pattern, increasing at the rate of about one centimeter per week, reaching thirty-eight centimeters at thirty-eight weeks.

However, this rule isn't hard and fast. Perhaps you noticed that a five-months-pregnant friend was about the same size as another expectant friend at seven months. Meanwhile, your belly may "pop" at six months. Height, weight, the number of previous pregnancies, and even ethnicity can contribute to this variability.

Don't worry if your belly isn't as big as your sister's or your friend's was at a certain point in her pregnancy, or if it isn't a perfect reflection of the fetal growth curve you see online or in a book. As far as "baby bumps" go, bigger doesn't always mean better.

A large study published by the National Institutes of Health (NIH) in 2015, in the *American Journal of Obstetrics and Gynecology*, confirms what many in the field have long suggested — that using a one-size-fits-all benchmark doesn't take into account the simple fact that women's bodies vary a lot. Researchers surveyed healthy, low-risk pregnancies of about seventeen hundred women of different races and ethnicities: Black, Hispanic, Asian, and white. Initially, the babies grew in utero at about the same pace, but after twenty weeks, differences began to emerge. At thirty-nine weeks, the babies of the white mothers were the largest, followed by those of Hispanic mothers, then Asian, then Black. (The differences were small, a matter of grams, but still evident.) Yet, *all* the babies and their mothers were considered low-risk and healthy. Adults come in all different shapes and sizes; babies do too.

Make Room for Twins (or More)

If the uterus can expand so dramatically for a single baby and placenta, imagine what it does when twins, triplets, or more take up residence! Nadya Suleman, the famous "Octomom" from reality TV, had not only eight babies but also multiple placentas.

It isn't always one placenta per baby, however. Sometimes siblings who cohabit the same uterus get a very early life-lesson in sharing. Identical twins develop from a single fertilized egg, and usually these babies share one placenta. The early-stage embryo splits in two during the first few days of life, making a carbon copy of itself (the reason that the babies look exactly alike). If this happens before placental cells develop, the babies will share the placenta. (The same basic principle holds regardless of how many times an embryo splits. If it cleaves into four pieces, then there will be four identical babies, and they will share the same placenta.)

Ideally the two (or more) siblings play nice and divide the placenta equally between them. When that happens, the babies will have similar weights. But imagine two people sharing an ice cream sundae — one dish, two spoons. Sometimes one person is hungrier or more determined than the other and eats more than half the sundae. That can happen with multiples sharing a single placenta. If one baby takes more than a fair share of the maternal resources, the other one (or more) may end up smaller.

Fraternal (non-identical) twins aren't forced into the same sharing arrangement. They each develop from their own fertilized egg, which in turn makes a placenta. Whether it's fraternal twins, triplets, or more, every baby gets its own.

However, as real estate agents like to say, location is everything — and in the case of multiple babies searching for nourishment in utero, the location of the placenta matters. Take the Octomom as an extreme example: with so many placentas in a uterus that was designed to hold only one, it becomes physically impossible for all of them to fit into "the sweet spot" at the top, where the roots of the placenta's tree grow best and the mother's nurturing blood is most plentiful.

In the contest to occupy this niche, there is usually a winner who edges out the competition. This baby weighs the most at birth. Suleman's babies were born at thirty weeks, so their overall weights were low. The heaviest came in at three pounds, four ounces, while the lightest was just one pound, eight ounces. You can probably figure out which baby was in the right place at the right time. Whether you're pregnant with a singleton or expecting more than one baby, it can get crowded in there.

HOW THE PLACENTA NOURISHES YOUR BABY

One way or another the placenta is going to provide your baby with the nutrients needed for growth. All aspects of its structure are optimized for efficient transport between mother and baby. However, it also utilizes numerous clever strategies to accomplish its goals, such as sending signals that mobilize a mother's food stores, which bolsters the level of nutrients in her blood that are available for transfer to the baby.

Before birth, your child is like the inhabitant of a space station, and you are the earth — Mother Earth, so to speak. Everything that is needed to sustain your baby's life in space (aka the womb) has to be transported from the earth (you). Your baby will require a state-of-the-art, highly efficient transportation system to get critical "supplies" from your body. The placenta is the go-between, a space shuttle of sorts.

Without functioning organs of its own, your developing child relies on the placenta to perform the work of multiple organ systems, which will take over once the baby is born. Like the lungs, the placenta exchanges oxygen for carbon dioxide. Like the digestive system, the placenta transfers food products to the baby, which fuel its growth and development. Like the kidneys, the placenta filters waste material.

Every item has its own transport mechanism. There are separate processes for transferring nutrients such as glucose, broken-down proteins (amino acids), fats, micronutrients (such as iron), vitamins (such as D), and water. Some are active, meaning that there are specialized conveyances that require energy, like minute versions of a freight train with specially designed cars to carry different types of cargo. Others are passive, meaning that substances just sail across on their own. The cells of the placenta are like a thin membrane between your blood and that of your baby: the two never mix. In some places, the two circulations are separated by a mere 1/10,000th of an inch. This super-thin wall promotes the passage of oxygen and small molecules from mother to baby by diffusion. They move from your blood, where they are more concentrated, to your baby's blood, where levels are lower. Once your baby has digested the nutrients, the leftover waste products, along with carbon dioxide, are transported in the opposite direction. The placenta returns them to your blood.

Scientists who study its transport properties think of the placenta as a kind of nutrient sensor that detects levels of these growth-promoting substances in maternal blood and adjusts their transport accordingly. It's like a supply-chain management system in the womb. For example, your baby requires vitamin D for bone growth, among other things. The placenta snares vitamin D in your blood and dispatches this essential nutrient to your baby; your overall vitamin D levels determine how much of this growth-promoting substance the placenta will extract and transport.

The placenta is a control point that matches your baby's growth to the amount of nourishing substances your body makes available. This means that if you are lacking enough of a key nutrient, your baby's growth may be restricted. This is why eating a healthy, balanced diet, supplemented with good-quality prenatal vitamins (discussed in the next section) as suggested by your doctor, is so important.

However, for reasons we don't fully understand, placental transport mechanisms sometimes don't work properly, compromising nutrient transfer to the baby, and growth becomes restricted. Conversely, an excess of nutrients can abnormally accelerate a baby's growth. Although not always the case, this can occur when mothers are diabetic or obese. Excess levels of circulating nutrients, such as glucose, stimulate placental transport, producing larger-than-normal babies.

Typically, though, if you are eating well and have a healthy supply of resources for your baby, the placenta will figure out the ideal supply-and-demand equation, transporting nourishment to your baby in the right quantities at the right time.

Prenatal Vitamins: D Is for "Do It"

Your placenta functions like your baby's in-house nutritionist. If you eat a well-balanced and healthy diet, do you really need to take specially formulated prenatal vitamins? The debate over their value can be noisy and confusing. *Yes, you have to take them for a healthy pregnancy . . . No, they're worthless and may even be harmful — just get your nutrition from food* . . . It's no wonder that many women question whether to take a prenatal vitamin supplement, before and during pregnancy. The answer is definitely yes.

One compelling reason involves vitamin D. This nutrient is best known for promoting bone health, preventing osteoporosis, and reducing the risk of fractures. Normal vitamin D levels also contribute to proper functioning of the immune system, helping to fight infection. They reduce the risk for cardiovascular disease and certain kinds of cancer.

In pregnancy, vitamin D has especially important roles, some of which are just beginning to be understood. Normal preconception levels are associated with higher pregnancy and birth rates while reducing losses. There are hints that continuing to take vitamin D may protect you against complications that are associated with abnormal placental structure, including preeclampsia. Supplementation with this nutrient also reduces the chances that a baby will be born underweight and improves postnatal growth. Vitamin D's positive effects on the immune system are in line with recent evidence suggesting that the children of mothers who take this nutrient during pregnancy have, for example, a lower risk of transient wheezing, an aftereffect of a viral respiratory infection.

Even if you're vigilant about your diet and aim for a healthy mix of foods that deliver plenty of nutrients, it's very difficult to get adequate amounts of vitamin D from what you eat. Foods that contain relatively high levels of vitamin D, including oily fish and fortified products like milk and cereal, make only minor contributions overall. By far the most significant source is sunlight, more specifically UV radiation, which stimulates production of pre–vitamin D in the skin. This precursor is stored in fat cells for later processing into its mature form, which our bodies can use.

Many factors — including desk jobs and indoor lifestyles, living in colder climates where we cover up to keep warm, plus the widespread use of sunscreens — have decreased our ability to produce the amount of vitamin D that we need to stay healthy. The CDC estimates that about 8 percent of people in the United States are deficient in vitamin D and that another 25 percent are at risk for inadequate levels. Vitamin D is particularly important if you are in a group with a heightened risk of deficiency, including women with dark skin and those who are obese or have had gastric bypass surgery. You are at even greater risk if you live in parts of the world where sunlight is at a premium.

The levels of vitamin D in a good-quality prenatal supplement are generally higher than what you'll find in an over-the-counter version, and that's likely true for some other nutrients that are essential for a healthy pregnancy, such as folic acid.

TAKE CHARGE: SUPPLEMENT WITH PRENATAL VITAMINS

Ask your doctor to recommend a high-quality prenatal vitamin, which may require a prescription. The right formulation is important. A recent study showed that mothers who, before and during pregnancy, took a multivitamin containing folic acid reduced their child's risk of developing autism spectrum disorder. Here's a list of nutrients essential to your health and your baby's. Through diet and supplements, aim to consume these daily levels:

• Vitamin D: 600 IU* (to support placental function and promote normal growth of the baby before and after birth)

• Folic acid: 0.4 mg in the first trimester, 0.6 mg in the second and third trimesters (to protect neural development)

• Iodine: 150 mcg (to prevent hypothyroidism in you and your baby)

• Calcium: 1,000 mg (to help maintain strong bones and teeth — yours and your baby's)

• Iron: 27 mg (to promote the production of additional red blood cells to deliver oxygen to the placenta and your baby)

Some pregnant women may experience nausea when they take supplements; contact your doctor if this happens to you. Switching brands or formulations may solve the problem.

The Placenta Regulates Maternal Metabolism

When you become pregnant, how your body uses food as fuel is utterly transformed. If you have a turkey sandwich and an apple for lunch when you're *not* pregnant, your body will metabolize those foods in a particular way that is focused on your needs, burning some for energy and putting the rest in storage for use later on. But if you have the same meal when you are pregnant, the metabolic process unfolds differently.

One of the placenta's most important jobs is to maximize the

---

* Vitamin D and some other nutrients are measured in international units (IU), a measure of activity rather than weight. Other nutrients in this list are measured by weight: milligrams (mg) or micrograms (mcg).

nutrients that are available to your baby, who needs ever-increasing amounts of these broken-down food products to sustain a normal rate of growth before birth. The placenta has many different strategies for provisioning your child. One tactic is to produce molecules that regulate hunger. It's not just your imagination telling you that you're hungrier when you are having a baby. The placenta is sending you powerful signals that increase appetite, so that you eat enough to feed both yourself and your growing child.

In a parallel strategy, the placenta mobilizes your body's stores of glucose, amino acids (the building blocks of proteins), and fats, raising blood levels of these nutrients and the amounts that are available to your baby. In the case of glucose, it goes to even greater lengths to make sure that your unborn child has the supply needed for normal growth. Products, including hormones, which the placenta releases into your circulation partially prevent your cells from taking up glucose, thus raising your blood levels so that the baby gets its share. Recently my colleagues and I discovered that placental products also cause the cells of the mother's pancreas to divide, increasing the organ's size and insulin production. This process raises blood levels of this hormone, compensating for the rise in glucose.

The amazing changes in maternal metabolism that pregnancy and the placenta bring about are another reason why it is important to eat regularly and well during pregnancy. If nutrient levels in your blood fall, the placenta ramps up its secretion of products that deplete your reserves of stored food, so that your baby can continue to receive nourishment. But this process can compromise your own nutrition.

TAKE CHARGE: HOW MUCH WEIGHT TO GAIN

Becoming pregnant doesn't really mean you're "eating for two." In fact, according to the American College of Obstetricians and Gynecologists, if your body mass index (BMI) is in the normal range, initially you should consume the same amount of food that you would if you weren't pregnant. For most women, that's about 1,800 calories a day (depending on activity level). When you reach your second trimester, you can add an additional 340 calories; in your third trimester, you need 450 calories more. The goal is a steady rate of weight gain.

The exact amount of calories depends on your BMI, which is easily determined. Use the internet calculator found on many websites, or apply this simple formula: *703 x weight in pounds / (height in inches x height in inches) = BMI.* For example, the calculation for someone who weighs 140 pounds and is 66 inches in height (5 feet 6 inches) looks like this: 703 x 140 / (66 x 66) = 22.6.

If you're of normal weight (with a BMI of 18.5 to 24.9), a rate of 0.8 to 1 pound per week is recommended, adding up to 25 to 35 pounds in all. If you're underweight (with a BMI of less than 18.5), the recommended rate is 1 to 1.3 pounds per week, for a total of 28 to 40 pounds.

Because we now know so much about the dangers (for both mother and baby) of being overweight or obese before and during pregnancy, women with higher BMIs should try to gain less. If you're in the overweight range (with a BMI of 25 to 29.9), you should limit your weight gain to 0.5 to 0.7 pounds per week, with a target of 15 to 25 pounds by the time of delivery. Obese women (with a BMI of 30 or greater) should aim lower, 0.4 to 0.6 pounds per week, for a total of 11 to 20 pounds.

HOW THE PLACENTA CHANGES YOUR BODY

We take for granted that pregnancy alters a woman's physical appearance — the growing belly and the enlarging breasts. But pregnancy also brings about other amazing transformations. Some, like morning sickness, are pretty obvious, but others, such as modifications in the way that your immune and vascular systems work, are not. All of these seemingly unrelated changes have one thing in common: they can be either directly or indirectly attributed to the placenta. The myriad products it releases into your blood affect how almost all parts of your body work.

Morning Sickness? Blame It on the Placenta

Not every pregnant woman experiences morning sickness, but about 80 percent do. Most cases are mild and can be dealt with by avoiding specific foods and situations, such as having an empty stomach, which can make things worse. At around week fourteen, nausea and vomiting usually subside (though a handful of women will, unfortunately, deal with this problem for an entire pregnancy). Researchers are still pondering why this is the case, but there's good reason to believe the placenta is involved.

According to an old theory — long debunked — morning sickness indicated that a woman felt ambiguous about her pregnancy. The controversial evolutionary biologist Margie Profet suggests that morning sickness is an ancient protective response — nature's way of getting our maternal ancestors to avoid poisoning themselves and their unborn babies with potentially toxic foods. We do know one thing for certain: the causes are genetic, not psychological. A family history of morning sickness raises your risk threefold.

Today research suggests that nausea and vomiting are triggered by the myriad products the placenta pours into the mother's bloodstream. One suspect is a growth factor (growth and differentiation factor 15, or GDF15), made in large quantities by the placenta and uterine cells. It's a case of guilt by association: GDF15 suppresses appetite, and a study showed that women who take anti-nausea medications during pregnancy have elevated levels of it in their blood. Also, GDF15 is implicated in causing symptoms similar to morning sickness in some cancer patients.

Another suspect is the hormone human chorionic gonadotropin, typically referred to by the acronym hCG. The first few placental cells that surround the embryo are already pumping out hCG in quantities so large, it can be detected in maternal blood and urine only days after implantation (which is why pregnancy tests are based on its presence).

The target of hCG is the ovary, which it holds in suspended animation, stopping the hormonal progression that normally leads to menstruation and the sloughing of the uterine lining. During pregnancy, this would lead to loss of the embryo. Women with morning sickness tend to have higher hCG levels in the blood than those who do not. The timing of its production coincides with the typical arc of morning sickness — rapidly rising early in the first trimester and declining in the second. As the placenta matures, it produces other hormones that prevent menstruation. The release of hCG wanes, and the levels of this hormone rapidly fall just as, for most women, the nausea and vomiting of pregnancy abate.

Can your morning sickness harm your developing baby? In most cases, the answer is no. However, 2 to 3 percent of pregnant women have intractable nausea and vomiting, a condition called hyperemesis gravidarum. It is important to let your doctor know if you suspect you have this severe form of morning sickness. One symptom is weight loss. If your weight drops by 5 percent, your doctor may order anti-nausea drugs. If these don't work, intravenous fluids and vitamins may be given to help support your baby's growth.

TAKE CHARGE: SNACK TO STAVE OFF MORNING SICKNESS

Although morning sickness can strike at any time of day, it's typical, as its name indicates, for the most acute symptoms to arise early in the morning. By then you haven't eaten for hours. Keeping a stash of crackers by your bedside and eating a few before you get up can help. Stepping up fluid intake, eating smaller meals more frequently, snacking on foods that are bland and easy to digest (such as bananas, rice, applesauce, toast, and tea, which are often referred to as the BRATT diet), and consuming foods made with real ginger (including soft drinks, tea, and candies) can be effective. If none of these strategies work, your doctor may recommend over-the-counter medications such as vitamin B6 and doxylamine (an antihistamine often used as a sleeping medication), alone or in combination, since both have been deemed safe for pregnant women. Simply avoiding smells that trigger the symptoms can be helpful.

Changes to Your Immune System

For decades, epidemiologists have gathered data showing that pregnant women are more susceptible to certain infectious diseases, such as the flu, malaria, and chicken pox. Similarly, pregnant women have an increased risk of getting certain foodborne illnesses, such as listeriosis, and toxoplasmosis, which is caused by a parasite commonly found in soiled cat litter. Besides making a mother-to-be quite sick, several of these disease-causing organisms can infect the placenta, causing damage and compromising its function. Some may also cross over to the baby, causing what is called a congenital infection, and do damage. Fortunately, with the exception of malaria, these diseases can be mostly prevented by making lifestyle changes and getting vaccines. (See chapter 1.)

Why pregnant women are more susceptible to certain infections is a question that has long captured the scientific imagination. This is partly because deeper knowledge of this phenomenon could have valuable applications for the transplant of organs, such as the kidney, liver, heart, or skin. Think about pregnancy as a unique form of organ transplantation: the baby is the donor of the placenta and fetal membranes, and the mother is the recipient. Then consider that half of the baby's cells come from its father — meaning that it's "half foreign" to the mother's body. In a typical transplant situation, such a mismatch would trigger organ rejection. It's remarkable that a baby growing in the uterus isn't recognized as "foreign" and destroyed by the mother's immune system. Somehow the placenta and fetal membranes form a protective shield that prevents this from happening.

Clearly, pregnancy is different from organ transplantation. Prior to any such transplant, a great deal of effort is put into matching certain characteristics of the donor and the recipient, such as size, blood type, and "transplant antigens," a set of molecules that govern rejection or acceptance of a transplanted organ. Even if a very good match is made, the recipient usually has to take powerful anti-rejection drugs to keep the immune system from attacking the organ that came from another individual.

None of these procedures are required for a successful pregnancy. You don't try to match blood types or inquire about transplant antigens with the future father of your child. Somehow the placenta and fetal membranes act as universal donors, making transplantation of any baby to the mother possible for the duration of pregnancy. Yet at the same time, a pregnant woman becomes more susceptible to certain infections. This evidence suggests that her immune system has been in some way suppressed, although the mechanisms that cause this are far from clear. The suppressed immune function allows a baby to grow in the uterus.

Further evidence, related to autoimmune diseases, supports the idea that pregnancy changes a mother's immune response. These diseases can be associated with an overactive or misdirected immune system. Symptoms of rheumatoid arthritis usually improve over the course of pregnancy; the same goes for multiple sclerosis: in both cases, a relapse typically occurs after delivery. Yet a simple down-regulation of the mother's immune system is not at work here, but rather something much more subtle. Consider these variations, based on the particular autoimmune disease: systemic lupus erythematosis patients may experience a worsening of the disease, which is one reason why they are counseled to attempt pregnancy when symptoms are at an ebb. Pregnancy has divergent effects on different forms of inflammatory bowel disease. Ulcerative colitis often gets worse. In fact, some women with this condition say that they don't have to get a pregnancy test — they can tell they are carrying a baby by the sudden worsening of their symptoms. In contrast, pregnancy seems to have little effect on Crohn's disease, neither exacerbating nor improving this condition.

Lots of theories exist as to why aspects of a woman's immune system seem to change during pregnancy. Decades ago researchers thought the very large hormonal shifts might play a role, a theory that fell out of favor but has recently been revived. Other scientists credit the special immune properties of the placenta or the uterus. Research is ongoing, but knowing what those mechanisms are — and figuring out how to reproduce them — will be a tremendous boon to the transplant field, where reducing the rates of tissue and organ rejection would help countless patients. There could be additional benefits for people with autoimmune conditions whose symptoms regress during pregnancy.

Widening of the Blood Vessels and Other Changes

The placenta changes how your vascular system works. The vessels through which your blood flows are like a highway. The only way to accommodate the approximately 50 percent increase in blood volume during pregnancy is to widen the road. The walls of your blood vessels "relax," increasing their diameter. This promotes the transport of more blood to the placenta and the delivery of nutrients and oxygen.

The placenta and the hormones of pregnancy bring about other remarkable changes, such as preparing your breasts for lactation. Like chorionic villi, the cells that make milk have a treelike structure. Starting early in pregnancy, placental products transform the simple branching pattern of the breast into a significantly enlarged network of much greater complexity. This is why breast size increases during pregnancy — a physical alteration followed by the complex molecular changes that are required for the eventual production of milk.

We know that the placenta is responsible for many of these changes because its delivery after birth rapidly reverses them. Aversions to certain foods and smells usually disappear quickly. Likewise, your risk of having a more serious case of the flu rapidly dissipates. When you are pregnant, particularly for the first time, it may seem hard to believe that eventually your body will return to normal, as if by magic.

But it isn't magic — it's simply the remarkable nature of a woman's anatomy.

THE PLACENTA MAKES US HUMAN

Why do we make such a big deal about the natural process of pregnancy and childbirth? After all, animals — including our closest primate cousins — routinely give birth with little fanfare and recover quickly to feed and care for their young. While we do have much in common with other mammals, especially our primate kin, human birth is biologically more complex, and the placenta is one reason why.

Evolutionary biologists mark the placenta's appearance as a major branch point of the evolutionary tree. Placental mammals thrived

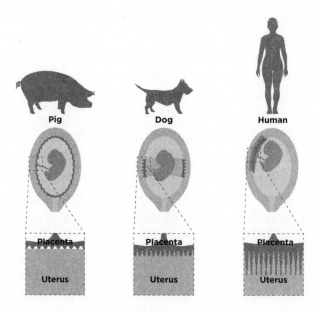

In many animals, such as pigs, the placenta spreads over much of the internal surface of the uterus. The placentas of dogs erode just enough of the uterine surface to lie next to its blood vessels. In humans, the placenta invades deeply into the uterus.

because they were able to channel a continuous supply of nutrients from the mother rather than from a yolk carried in an egg, a decidedly limited source that can support development for just so long. This adaptation was one of the seminal events that made us human.

Usually, the major organs of mammals have similar structures. The human heart and the cow heart, for instance, look much the same inside and out, and even work similarly — four chambers separated by valves, which open and close as blood is pumped throughout the body. But this is not true for the placenta. Its structure is highly variable among placental mammals.

The placentas of many farm animals lie on the uterine surface. The trophoblasts don't travel deep into the uterus and don't reroute blood to the placenta. Like a mossy covering that's easily removed from the soil, the placentas of these animals are easily separated from the uterus. They simply peel off. Placentation is a bit more invasive in dogs and cats, but not all that much. With us, it's a very different story. The roots of the placental tree dive deep into the uterine wall, where they form elaborate connections with the mother's blood vessels. Very few animals have this type of deep placentation. It's logical that monkeys, our close neighbors on the evolutionary tree, might share some aspects of human placental structure, but surprisingly, guinea pigs do as well.

Why is placental structure so different among animals? Many scientists think that the aggressive way in which the roots of the placental tree seek nourishment from the uterus enabled the evolution of the human brain, which requires an enormous investment of maternal resources in the form of the food it needs to develop and wire its sophisticated structure.

However, there is a downside to deep placentation: it makes removal at birth a tricky affair. Imagine pulling a plant out of the ground with its highly developed root system intact. If you pull gently and it comes out whole, the earth fills in the space where the plant once grew. Human birth ends in a similar way. Uterine contractions expel the placental roots from the uterus, which shrinks to fill in the space they once occupied. In the process, the blood vessels that once fed the placenta clamp shut. Sometimes, for reasons researchers don't fully understand, these uterine blood vessels don't constrict as they should, resulting in bleeding during or following delivery, a serious development that requires immediate medical intervention. It's one of many good reasons to have your baby in the hospital.

3

Understanding Your Prenatal Testing Options

N ot all that long ago, a child's life in the womb was shrouded in mystery. The visible changes in a mother's body were the only evidence that pregnancy was progressing: the swelling of the belly, the faintly discernible movement of the growing baby starting in the second trimester, followed by the unannounced magic show that the tiny being put on as it stretched within the increasingly crowded confines of the uterus, revealing an occasional suggestion of a foot or the thrust of a pointy little elbow. After months of wondering what was going on, these brief sightings offered exciting proof of an actual baby. Girl or boy? Two eyes and two ears? There was no way of knowing until the moment of birth.

Today, modern obstetrical care includes high-resolution photo documentation of your baby growing in the womb. You can get 3D, 4D, or even 5D ultrasound pictures of your child to take home and share with friends and family. Placental structure and function are also coming into clearer focus, thanks to science and technology. And beyond the advances in prenatal testing, there is also the burgeoning field of fetal surgery, where it's possible to correct life-threatening developmental defects and genetic diseases *before* a baby is born.

Aside from the routine testing that your prenatal health-care provider will recommend, the array of screening and diagnostic tools available today can be confusing. You may be wondering how to evaluate the latest options and whether the newest procedure will be beneficial for you and your baby.

Whatever tests you choose, understanding their objectives, their underlying technologies, and their pros and cons is essential for taking charge of your pregnancy. While much of the *what's-going-on-in-there?* mystery is gone in a twenty-first-century pregnancy, it's been replaced by a wealth of valuable, science-based information. It can empower you to make critical decisions that can improve your chances of having a healthy pregnancy and baby.

WHAT CAN TESTING TELL YOU?

Before we sort through the various screening and diagnostic options, let's look at the information that you can receive from prenatal genetic testing. (For a discussion of carrier screening for you and your partner prior to pregnancy, see chapter 1.)

Human DNA is organized into tightly packed bundles called chromosomes. Normally a human cell has twenty-three pairs, forty-six in total. One chromosome in each pair is from the mother and the other is from the father. Twenty-two of those pairs are identical in males and females; they contain directions for making and maintaining the parts of our bodies that both sexes have in common — the heart, eyes, skin, limbs, and so forth. The twenty-third pair consists of the sex chromosomes, which control whether your baby is male (XY) or female (XX).

The illustration on page 71 depicts a normal "karyotype," or picture of an individual's chromosomes. Some pairs are short and some are long. In addition to differences in length, each pair has a unique banding pattern visible with special dyes: horizontal stripes in different places because each carries a unique package of DNA. Paired chromosomes are mirror images of each other, but because one chromosome comes from each parent, they may contain different versions of the same genes.

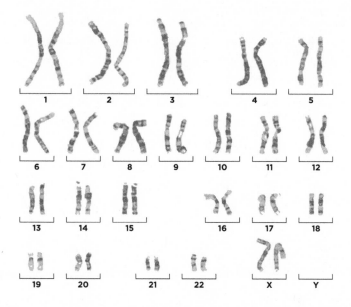

Chromosome pairs in a standard karyotype such as this one are arranged from longest (number 1) to shortest (number 22). The twenty-third pair consists of the sex chromosomes, XX in this example, which identifies this individual as female. A male has an XY pair.

As a woman ages, her chances of producing a genetically abnormal (or aneuploid) egg naturally increase. Previously, if you knew you would be thirty-five years of age or older on your due date, your health care provider would recommend prenatal genetic testing. However, most babies with genetic disorders, such as Down syndrome, are born to women age thirty-five *and under* — that's because they have far more babies than the over-thirty-five age group does. In other words, prenatal genetic testing isn't just important for women deemed high-risk due to their age; it can be valuable for all pregnant women, no matter how old.

The most common problem with a genetically abnormal egg is an extra copy of an entire chromosome. If this aneuploid egg is fertilized, a genetic disorder caused by the incorrect number of chromosomes will be passed on to the embryo. As a result, the baby that develops

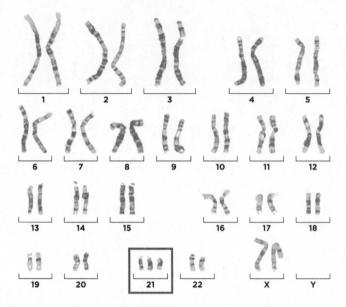

These results of a prenatal genetic test show three, instead of two, copies of chromosome number 21, indicating Down syndrome.

has three copies of a particular chromosome instead of two, a condition that geneticists call trisomy.

Trisomy most often affects chromosome 21, causing Down syndrome. The karyotype is shown in the figure above. Less frequently, there are three copies of chromosome 18 (Edwards syndrome) or chromosome 13 (Patau syndrome). These are rare disorders, and the affected babies usually don't survive. Trisomies involving other chromosomes are relatively uncommon; when they occur, the pregnancy usually ends in early miscarriage. A baby may also carry too few or too many copies of a sex chromosome, typically X rather than Y.

A baby with a common trisomic disorder will likely have associated developmental defects, such as abnormalities in the formation of the brain, the heart, and other organs. The placenta is also one of the baby's organs, and as such, it too will be affected by the genetic disorder. Because these conditions unfold as the baby and its placenta develop in the uterus, aneuploid pregnancies are sometimes more susceptible to loss or complications such as preterm labor or preeclampsia (discussed in chapter 5).

Prenatal genetic testing analyzes fetal chromosomes or DNA obtained through various methods, looking for abnormalities such as an extra chromosome. Amniocentesis and chorionic villus sampling (known as CVS) were once considered the go-to options for mothers in the thirty-five-and-up group who wanted to know if their babies would be born with Down syndrome or other genetic disorders. Though extremely accurate and valuable as diagnostic tools, they are both invasive procedures that carry a small but real risk of miscarriage. Amniocentesis, done between weeks fifteen and twenty, involves the withdrawal of amniotic fluid and the fetal cells it contains. CVS, performed between weeks ten and twelve, obtains a tiny bit of the baby's placenta.

Under certain circumstances, both procedures — discussed in more detail later in this chapter — may still play a vital role in providing critical information on a baby's health. But in recent years, they've been supplanted by a new approach that's rapidly becoming the method of choice: cell-free fetal DNA testing, also referred to as NIPT (non-invasive prenatal testing). This screening tool emerged from the unexpected discovery that fetal DNA circulates in a mother's blood. Now the risk of a genetic disorder, including Down syndrome, can be estimated as early as ten weeks by a test that is essentially risk-free to the baby because it's done with a blood sample taken from the mother.

If you've had a prenatal visit with your health care provider, you've likely come across information on cell-free fetal DNA testing. It's the latest thing — but is it right for you and your baby? It depends on a number of factors, so before you decide, you'll benefit from learning more about NIPT, including its advantages and its limitations.

THE CELL-FREE FETAL DNA REVOLUTION

Early in my career I became intrigued by the idea that placental cells might not stay put. Because a mother's blood sprays like a fire hose on the placenta (see page 46), it's easy to imagine how these jets might dislodge cells from the placenta, launching them into the mother's bloodstream like little boats carrying the baby's genetic information.

I was certainly not the only one to become intrigued by this idea and its potential benefits, including a prenatal genetic blood test, with essentially zero risk, which could be offered to all pregnant women. Several major funders stepped forward with support for researchers exploring this concept. Our group worked on trying to fish out placental cells from maternal blood, and other research teams targeted fetal red blood cells. Neither approach was robust enough to be widely adopted. Nevertheless, I continued to closely follow research in this area.

And then a ray of hope appeared. Presenting his results at a conference that I attended, Dr. Dennis Lo, a professor of chemical pathology, suggested that substantial amounts of "cell-free" fetal DNA circulated in maternal blood. It's described as "cell-free" because somehow this DNA exists outside fetal cells rather than bound up in their nuclei. Dr. Lo was questioned closely, and his findings were vigorously debated. Many attendees found his results to be nearly (or entirely) inconceivable.

Now, many years later, Dr. Lo's conclusions have been confirmed. The placenta pours cell-free fetal DNA into maternal blood, which can be used for prenatal genetic testing — and there is so much DNA present that, in most cases, a conventional blood sample is all that is needed. It is not surprising that this nearly risk-free procedure is rapidly eclipsing the old tests, and analysis of this cell-free fetal DNA is revolutionizing prenatal genetic testing.

The blood sample obtained for NIPT is scrutinized by one of several competing technologies. No matter how it's done, the ultimate goal — as with amniocentesis and CVS — is to analyze fetal DNA and inventory the chromosomes to check for an abnormal count indicative of a genetic condition.

Today millions of pregnant women all over the world are having their blood drawn for cell-free fetal DNA testing. If you're wondering whether to pursue this test, the first thing to know is this: cell-free fetal DNA testing is a screening tool, not a diagnostic test. Knowing the lingo — and the difference between these two types of tests — is important.

> **TAKE CHARGE: SCREENING TESTS VS. DIAGNOSTIC TESTS**
>
> Descriptions of medical tests are rife with terms that are easy to misunderstand because they are often used interchangeably in general conversation. A good example is "screening test" versus "diagnostic test," two terms that have very different meanings. In clinical practice, a screening test is used to identify a person with an elevated risk of having a particular condition. To confirm the presence of that condition, a second test, the definitive diagnostic one, needs to be performed. Medical decisions concerning surgical interventions or drug treatments are based on diagnostic tests.
>
> When you become pregnant, you'll hear about lots of testing options for you and your baby. Listen for the words "screening" and "diagnostic"; they indicate the level of information you can expect from test results. (And if your provider uses any medical term you do not understand, ask for an explanation.)

Is Cell-Free Fetal DNA Testing Right for You?

The medically safe nature of NIPT is a major reason why the American College of Obstetricians and Gynecologists now recommends that women of all ages, regardless of whether their risk is high or low, be offered genetic testing before the twentieth week of pregnancy.

The high-risk group includes mothers who are thirty-five years of age and older. A positive result from another screening test, such as the multiple marker and ultrasound tests described later in this chapter, will also put you in this group. Less common reasons for being considered high-risk include a prior pregnancy in which the fetus was trisomic and the rare situation in which a parent is known to have a "chromosomal translocation," meaning that large portions of genetic material from two *unpaired* chromosomes are swapped, making new hybrids that may not function correctly.

If you are in a low-risk category, for example, under the age of thirty-five, you will be offered the option of NIPT. In the United States, whether you will get this test is largely determined by your insurance company's willingness to reimburse the costs or your ability to pay out of pocket.

Here's what the results of NIPT can tell you. It's an effective screening test for trisomy 21 (Down syndrome) and only slightly less so for trisomy 18 (Edwards syndrome), but it has relatively poor predictive value for trisomy 13 (Patau syndrome). If you receive a positive result indicating a heightened risk of a trisomy, remember that this is not a firm diagnosis. Instead, a positive result alerts your clinician to the need for another screening test or a procedure that *will* give a definitive result — either chorionic villus sampling or amniocentesis.

A positive cell-free fetal DNA screening test triggers a series of events that take time — the time needed to schedule another test and the time needed to determine who will pay for it (you, your insurer, or possibly the state in which you reside). It also takes time to obtain the results from the follow-up tests. As a result, women and their families have to endure an excruciating period of uncertainty. This is why false-positive test results are such a problem when it comes to prenatal genetic diagnoses. Advising a woman that she might be carrying a child who has a trisomy causes an enormous amount of unnecessary stress — if later, it turns out not to be the case.

Unfortunately, a cell-free fetal DNA test does not always yield an informative result, as happens to 1 to 5 percent of women. There are several possible causes. One is obesity (defined as a body mass index of over 30), in part because obese women have higher amounts of their own DNA in circulation, obscuring the signal from fetal DNA. Another cause is related to the fact that the fetal DNA is coming from the placenta. In a small number of cases, the placenta is like a patchwork in which some parts are genetically normal and others are abnormal, a situation that produces a confusing result. The testing is also less conclusive for mothers carrying twins or other multiples.

Finally, it is possible that an abnormal genetic signal may be coming from the mother rather than the fetus. For example, NIPT may turn up evidence that some of her cells have lost one of the two copies of the X chromosome. These tests are teaching us that variations in chromosome numbers are more commonly detected in normal women than was once thought, a very surprising result.

Whether or not you choose to have this test is, of course, a personal decision, but you should start by considering where you may fall on the risk scale, including your age, any known family genetic disorders, and your history, if any, with prior pregnancies. Discuss your concerns with your health care provider.

Some women are having the test done for its intended purpose: as a prenatal genetic screening tool. Others use it primarily to learn the sex of the baby early in pregnancy. However, it is possible that those who opt for cell-free fetal DNA testing just to know their baby's sex will get more information than they bargained for. Women in the low-risk group who want reassurance that their unborn baby is developing normally can consider other types of lower-cost screening tests, discussed below. These proven methods have been used for decades.

MULTIPLE MARKER TESTING

Like analysis of cell-free fetal DNA in maternal blood, multiple marker testing uses a blood sample. But unlike DNA tests, these screening options are relatively inexpensive and therefore more widely in use.

The "marker" approach to testing began in the UK during the 1970s, following the discovery that maternal blood levels of alpha-feto protein (AFP), produced by the baby and placenta, were higher when a fetus had a major defect involving the brain or spinal cord, such as spina bifida. (The spinal cord is exposed through an opening in the back.) For the first time the risk that an unborn child might have a serious problem of this sort could be quantified. If the amount of AFP in your blood was above a certain level, there was a high likelihood that you would give birth to a baby with these problems. Fortunately, scientists discovered that AFP screening, in combination with folic acid supplementation before or at the time of conception, could prevent about 80 percent of these defects. (For recommendations on folic acid supplementation, see chapter 2.)

Eventually, the single marker test was made more specific by including the levels of up to three other molecules produced by the baby or the placenta. In addition to AFP, the multiple marker test assays levels of the hormones hCG (human chorionic gonadotropin), estriol, and inhibin A, all critical indicators of fetal development.

These days the quad screen (a term that is sometimes used interchangeably with "multiple marker screening" or "maternal serum screen") is considered the test of choice for all pregnant women. The quad screen can be done between weeks fifteen and twenty of a pregnancy.

THE INTEGRATED TEST

In the early 1990s, as ultrasound methods for obtaining live images of babies advanced, a new measurement called nuchal translucency, used in conjunction with the marker tests done with a sample of maternal blood, came into being. By observing ultrasound images taken near the end of the first trimester, experts were able to link the apparent accumulation of an abnormally large amount of fluid at the back of the baby's neck with an elevated risk of Down syndrome or Edwards syndrome, as well as certain heart defects. The ultrasound measurements were incorporated into the formulas used to estimate a woman's risk of having a baby with one of these trisomies. Because it melds two types of data, the results of a marker test and an ultrasound image, this screening tool is known as the integrated test.

MIX AND MATCH: WHICH SCREENINGS TO GET

Today different versions of prenatal tests use various combinations of marker levels and ultrasound findings as screening tools to estimate a pregnant woman's risk of carrying a baby with a trisomic disorder. These tests can also pick up other types of developmental problems, such as major malformations of the brain or spinal cord.

Talk with your health care provider to determine which screening tests are right for you and when you should have them. You'll also want to find out which ones are covered fully or in part by your insurer, and if you qualify for state-based programs that pay for some or all of the costs.

In general, the screening tests described here are done in the first and second trimesters, but because every pregnancy is different, there is not an exact calendar or combination of testing options that all women should follow to the letter. One way to approach your decisions, besides considering your health history, is to focus on your age:

* *If you are under thirty-five and are considered low-risk,* you have various options. For instance, if cell-free fetal DNA testing in the first trimester indicates low risk, its accuracy can be boosted by adding the results of a blood test for assessing marker levels done in the second trimester. In another scenario, if the results of a quad screen after week fifteen indicate a potential issue, the next step may be ultrasound, perhaps followed by amniocentesis for a diagnosis.
* *If you are thirty-five or over or have a family history of genetic disorders,* you still have numerous options, but your health care provider may suggest that you move directly to diagnostic testing (amniocentesis or CVS) instead of screening, though it's entirely up to you.

When it comes to prenatal genetic screening tests, you have choices, and understanding the tests themselves can help you make an informed decision. Now let's take a closer look at the most common forms of diagnostic testing.

AMNIOCENTESIS AND CVS

If you are pregnant with your first child and will be thirty-five years of age or older when your baby is born, you're in good company. There are lots of great reasons to wait until you're ready to become a mother, ranging from having a healthy relationship with your partner and a stable home life to feeling secure in your career to simply being emotionally ready to take the life-changing plunge into parenthood! For more and more women, that sweet spot is the mid-to-late thirties. From 2000 to 2014, the number of first-time mothers ages thirty to thirty-four rose 28 percent, and the number of first-time mothers ages thirty-five and over rose 23 percent. According to the CDC this trend has continued in subsequent years.

You may also be thirty-five or older and pregnant for the second or third time. Whether or not this is your first baby, once you enter this age group, as discussed earlier, the chances of producing an egg with an extra chromosome or other abnormality increase. Though NIPT has revolutionized genetic screening, it is not a substitute for diagnostic testing. And while the quad screen and the integrated test can provide important information about your baby's health, only CVS (performed between weeks ten and twelve of a pregnancy) or amniocentesis (performed between weeks fifteen and twenty) can provide an actual diagnosis if screening tests indicate the possibility of a genetic condition.

These tests are the most direct route to assessing your baby's genetic health because his or her own cells are analyzed. The tests are, however, invasive and carry a measurable risk of miscarriage — from about 0.5 to 2 percent. The needle or tubing used in these procedures may accidentally damage a blood vessel in the placenta, which can, in a small number of cases, cause significant bleeding. Other complications include infection or rupture of the amniotic sac.

Given these risks, how does an obstetric care provider decide to recommend invasive testing? Advice is based on a risk-to-benefit ratio:
at age thirty-five a woman's chances of having a child with a genetic
abnormality become approximately equal to her chances of losing a
baby due to amniocentesis or chorionic villus sampling.

In amniocentesis, a needle is inserted through the abdomen to
remove a small amount of amniotic fluid and the fetal cells it contains. In CVS, a tiny piece of the baby's placenta is obtained, either
with a thin piece of tubing inserted through the vagina and cervix, or
with a needle passed through the abdomen. The general consensus
among women who've experienced these procedures is that they are
painless, though CVS sometimes causes a bit of discomfort. But there
is no need for an anesthetic with either test.

With both amniocentesis and CVS, the extracted samples are used
to grow cells in a laboratory for genetic analyses. It is easy for experts,
using high magnification, to count the number of chromosomes that
the growing cells have. They're generally searching for an extra copy
of one of the troublemakers — most commonly chromosome 21 for
Down syndrome — but their work can also be guided by the results
of a screening test that suggests a specific disorder.

Because of the risk of miscarriage, the decision to go through with
amniocentesis or CVS can be an agonizing one for women and their
families. The rule of thumb in searching for the right provider to
administer the test is that the more procedures that a provider has
done, the lower the risk to the patient — in this case, mother and baby.
Practice makes perfect. Still, it's not always easy to find an obstetric
care provider with this skill set (in part because invasive tests have
been eclipsed by NIPT). Find out how much experience your clinician has before signing up for the procedure, and don't hesitate to
ask about the person's track record. If you think that you may be a
candidate for these tests, discuss them with your health care provider
early on. If you need to line up an experienced professional for these
procedures, you don't want to attempt doing so at the last minute.

The other factor, of course, that makes this decision hard is that
you may be faced with a serious diagnosis that requires action, such

as a pregnancy termination, within a certain time frame. Or you may choose no action, and acceptance. In any event, the key word is *choice*, and the choice is yours.

If You Want More Information

If fetal ultrasound tests (discussed in the next section) turn up a birth defect, such as a heart problem, you might consider a chromosome microarray analysis, which the American College of Obstetricians and Gynecologists recommends. This test, which is done on samples obtained by amniocentesis or CVS, is an example of a highly specialized genetic technology used for diagnostic purposes. Rather than looking for an extra third copy of an entire chromosome, this method is akin to taking a magnifying glass to each chromosome to look for small-scale mistakes. Because of its high resolution, this technique gives a great deal more information than the traditional chromosome counting methods — something that it also does.

Being able to attribute a structural problem to a known genetic condition helps answer critical questions and clarify the clinical course. Can action be taken to correct the defect? Are there other serious problems that are not visible by means of ultrasound?

But the strength of this technology is also a weakness. If we take a magnifying glass to the DNA of *any* ostensibly normal individual, we may find tiny mistakes. Sometimes we know what they mean. A microarray analysis, for instance, could reveal an increased risk for cancer later in life, or another adult-onset disease. As a parent, you will be living with this knowledge about your child's health and will eventually have to decide whether and when to pass this information on.

Another possible outcome is a murky result, called a variant of uncertain significance (VUS). The name denotes the problem. We don't yet know if the change is a normal variation or may cause the baby to develop a health issue in the future. This can be sorted out by testing the parents. If the variant is traced to one or both parents with a normal health history, then there is a high likelihood that the change is inconsequential. If the VUS appears only in the child, then it's a "wait and see" situation.

The results of chromosome microarray analysis are more complex than the yes-or-no answers usually gleaned from the screening and diagnostic tests we have discussed. For this reason, if you are going to proceed with this analysis, clinicians strongly recommend comprehensive pre- and post-test genetic counseling with an expert in this technology so that you and your partner know what to expect and how to interpret the results.

FETAL SURGERY: RARE, BRILLIANT, AND PROMISING

Only a handful of pregnant women and their babies have undergone fetal surgery — surgical interventions performed in utero — but it is a groundbreaking branch of medicine that could someday impact a family's prenatal testing decisions. At my institution, Dr. Michael Harrison pioneered in 1981 a procedure to correct a structural defect in a baby's urinary tract. An ultrasound detected that the baby's blocked kidneys had filled with urine, expanding like water balloons. The kidneys were destined to fail. Dr. Harrison performed a surgery involving the insertion of a small tube in the bladder, which allowed the urine to escape. That baby is now an adult. Other still-rare candidates for fetal surgeries include cases of spina bifida. Closing an abnormal opening in the back and spinal cord may somewhat improve a child's neurological function. A few heart and lung defects can also be repaired by fetal surgery. In working out these techniques, researchers made a surprising discovery: a fetus heals with minimal scarring or even none at all. If ways could be found to make the surgeries safer, then other non-life-threatening conditions, such as cleft lip, could be corrected before birth with a better result.

More recently, my colleague Dr. Tippi MacKenzie and her team successfully transplanted blood stem cells from a pregnant mother, who was a carrier of the genetic blood disorder alpha thalassemia, to her unborn affected daughter, who inherited copies of the disease-causing form of these genes from both parents, resulting in alpha thalassemia major. It's probable the baby would have died before or shortly after birth, with severe organ damage due to anemia and the associated lack of oxygen in the blood. The stem cell treatment, which consisted of blood transfusions through the umbilical cord, was part of a clinical trial, and the child will likely require a bone marrow transplant to cure the disease. Still, fetal surgery enabled doctors to begin treatment of this usually fatal illness *before* birth. Researchers are hopeful that this type of in utero stem cell transplantation could one day help treat babies diagnosed with the more common form of thalassemia (beta thalassemia), sickle cell anemia, and other blood disorders.

Fetal surgical interventions are anything but routine and are risky for both mother and baby — but there's no question that their life-changing potential is enormous.

PRENATAL GENETIC TESTING: THE CHOICE IS YOURS

Testing is not mandatory — it really is *your* choice, as is the decision to opt for screening versus diagnostic testing. Your clinician will outline your options and likely make a recommendation, but you should never feel forced into choosing one test over another or testing in general.

Still, in a busy clinical environment, it's very possible that you will not be given all the information you need to make an informed choice, particularly if timing is an issue. For instance, you may have only a few weeks — or even days — to decide whether to have a particular test because the window is closing on when termination is possible. However, armed with the information about prenatal tests summarized in this chapter, it may be easier for you to know which questions to ask your clinician.

There are many factors that favor some form of prenatal testing. For most women under the age of thirty-five, the optimal strategy is multiple marker screening, with the option of diagnostic testing if screening results come back positive. For the thirty-five-and-over group, experts recommend cell-free fetal DNA testing as a primary screen and the optimal starting point for genetic testing.

The Society for Maternal Fetal Medicine does not recommend that all pregnant women be offered cell-free fetal DNA screening or that insurance companies should pay for it. But they also state that for ethical reasons a low-risk patient who has been informed of the facts and still wants the test should be allowed to have it.

Most women receive negative test results, which bring valuable assurance and peace of mind. However, the screening tests are not perfect. There is a small chance that you will receive false negative results, meaning that the baby has a chromosomal problem that was not picked up. Negative results are reassuring, but not proof positive.

Before you choose whether to have prenatal genetic testing, it's important to ask yourself the following question: what will you do with the results? If this information will not change what you are planning to do, then it may not be worth having such a prenatal test. On the other hand, the results may still be valuable because the information can be used to optimize management of your pregnancy.

ULTRASOUND: A CLOSE-UP VIEW OF YOUR BABY

Because ultrasound during pregnancy is now considered routine, we tend to take this technology for granted, but it's a relatively recent invention. In the United States it was introduced in the late 1970s and early 1980s, and the technology used in that era was not nearly as sophisticated as what we have today. It's possible that your own mother had such an ultrasound, craning her neck to look at a screen that showed a very grainy, murky, black-and-white "picture" of you in utero.

Like the evolution of television from its early days to the present, the ultrasound picture has improved considerably in the twenty-first century — sharper, bigger, clearer, with dimension and depth, and in some instances, images have made the leap from black-and-white to color. Doctors, researchers, and technicians (including the trained sonographers who do this procedure) are also more knowledgeable than ever about what those pictures can tell us, as well as what their limitations are.

The basic principle behind ultrasound imaging is akin to what goes on when you see yourself in a mirror. Light bouncing off your body hits the mirror and is reflected back as your image. Ultrasound works in a very similar way. Sound waves that are passed through your abdomen reach the baby and are reflected back, producing echoes that a computer turns into images with an extraordinary amount of detail.

OB/GYNs and their radiology colleagues consider ultrasound technology to be quite safe for both mother and baby. This makes sense — the pictures are produced by echoing sound waves, which are thought to be harmless. In contrast, X-rays are a fundamentally different technology. They use ionizing radiation, dangerous in high doses. Nevertheless, OB/GYNs recommend that ultrasound imaging be done only for a medical reason (as in the case of routine second-trimester screening) and not just for the fun of seeing the baby.

The Second-Trimester Ultrasound Scan

As part of standard obstetrical care, the first pictures of your baby can be taken many weeks before birth. The purpose of a second-trimester ultrasound, typically done between eighteen and twenty-two weeks, is to make sure that development is normal and that everything is in the right place.

Most major medical organizations and a range of pregnancy experts recommend that you have at least one scan during pregnancy, and the second trimester is the optimal time because the baby has developed to the point where various organ systems, physical features, and their functions can be assessed visually by a trained clinician. In some cases, ultrasound may be recommended in the first trimester (discussed later in this chapter). Should you have a first-trimester ultrasound, it is often done transvaginally, rather than transabdominally, the scanning discussed here. That's because the baby and its uterine home are still in close proximity to the vagina, having not yet ballooned upward. However, a transvaginal ultrasound may also be done later in pregnancy to assess the length of the cervix, the tunnel-like passageway that opens into the vagina and shortens in preterm labor.

Like prenatal genetic testing, ultrasound scanning is optional, but experts agree that it's important to do. That's because most women who have babies with the sorts of problems detected by ultrasound (such as low amniotic fluid or a heart condition) have no risk factors that otherwise would cause concern.

The exam has several components. One aspect focuses on fetal development. An initial assessment determines how many babies there are, confirms the presence of two hands and two feet, and checks whether the heart is functioning normally. It should be beating at regular intervals between 120 and 160 times per minute. If this is your first pregnancy and you have yet to hear your baby's heartbeat, be prepared for a very rapid rhythm! Many parents are surprised at how quickly that tiny heart beats, but this is perfectly normal.

All parts of the baby's body are surveyed in detail. Starting with the head, normal brain anatomy is confirmed and the upper lip and nose are imaged to rule out the possibility of a cleft, or the failure of the right and left sides to join in the middle.

If you have had multiple marker or a cell-free DNA screening test, examining the back of the neck for an abnormal thickening, discussed above, can increase the accuracy of estimating the risk of a genetic abnormality. But this step is skipped if you've had amniocentesis or CVS, which are diagnostic procedures.

In the baby's chest area, the sonographer peers inside the heart to make sure that the four chambers and the big vessels that carry blood to the lungs and the rest of the body are arranged correctly. The abdominal organs are examined to verify that all parts of the digestive and urinary systems are forming properly. The back is scrutinized to make certain that its interlocking bones are encasing the spinal cord, which by now should be shaped like a rod.

Another important result that the scan produces is a more accurate measurement of gestational age (the age of your pregnancy) and, therefore, your due date. As you likely know, your clinician calculates your due date from the first day of your last menstrual period (LMP) by adding 280 days. Given wide variations in the timing of ovulation and the irregularity of many women's cycles, the LMP dating method can miss the mark. If you conceived through in vitro fertilization, gestational age is calculated based on the age of the embryo and the date of transfer, which can be a more precise method.

The second-trimester ultrasound either confirms the original gestational age and due date or enables a more accurate revision. This is done through a combination of measurements that include the length of an imaginary straight line drawn between the baby's ears, the circumference of the head, the circumference of the abdomen, and the length of the thigh bone.

For many parents, one of the most exciting things about the second-trimester ultrasound is the ability to determine if the baby is a girl or a boy. A first-trimester scan does not have this discriminating power. In the second trimester, the anatomical differences are usually plain to see. If you don't wish to know the sex of your baby in advance of birth, make sure to tell the clinician who is doing the ultrasound. He or she may assume that you want that information. And if you or your partner is particularly eagle-eyed, don't be too surprised if you figure it out on your own. (There are certainly stories of parents who decided they were having a boy or girl based on what they *thought* they saw — or didn't see — only to be caught completely off guard by the great reveal at birth!)

In addition to examining the baby's development and anatomy, the second-trimester scan gives important information about the placenta, including its location. Is it attached away from the cervix? Or does it meet the criteria for placenta previa, covering a portion or all of the cervix? It's important to have this information before you give birth, as you know from chapter 2.

The ultrasound images also allow the sonographer to estimate the amount of amniotic fluid that surrounds the baby. At eighteen to twenty weeks, the average volume is a little less than two cups. Too much (called polyhydramnios) or too little (called oligohydramnios) may signal that the baby is having problems. Why? By this stage of pregnancy the baby is normally drinking amniotic fluid. The effect is the same as when you drink water. The kidneys turn excess liquid into urine, which is excreted into amniotic fluid, contributing a substantial amount to the volume. A disruption in any part of this cycle can lead to an abnormal amount of amniotic fluid.

The second-trimester ultrasound scan includes an evaluation of the ovaries (looking for cysts or tumors), the uterus, and the cervix. The length of the cervix is important, as cervical shortening is associated with preterm birth (discussed in chapter 5). Finding out that you have this condition, which can occur without any obvious signs, will allow your clinician to recommend next steps to prevent further shortening, a possible prelude to cervical dilation and the initiation of labor, which may be impossible to stop.

Obesity can alter how second-trimester scans are done. When performed on obese women in the routine way, an ultrasound is about 20 percent less capable of detecting the structural problems it normally captures. In such instances, to get more accurate information clinicians may use a transvaginal scan to obtain images. The sound waves travel to the fetus through the vagina rather than the abdominal wall. Another strategy is to delay the scan until later in the second trimester, when fetal growth makes it easier to see the structures that are routinely examined.

Beyond Routine Ultrasound Scans

Your practitioner might suggest an ultrasound at a time other than the standard eighteen to twenty-two weeks. There can be good reasons for doing a scan as early as the first trimester or much later, in the third trimester.

Before the fourteenth week of pregnancy it's too early to see many of the fetal body parts the test was designed to image. It's also hard to reliably determine the baby's sex in those early weeks. But a first-trimester ultrasound is effective at estimating gestational age. A clinician measures the distance from the top of the baby's head to the tip of its bottom, which is called the crown-rump length. The length can be plotted on a graph that accurately converts this measurement to gestational age. Evidence that a baby is growing normally is powerful reassurance that a pregnancy is on track.

A first-trimester ultrasound is also a good way to determine if the placenta is implanted inside the uterus or outside its walls. Implantation in a Fallopian (uterine) tube, for example, will produce an ectopic pregnancy (see chapter 2), which requires immediate action. Alternatively, a first-trimester scan may be done for a particular reason, such as confirming that the heart is beating, which can be detected as early as six weeks by ultrasound. In certain cases, genetic disorders can be detected by a first-trimester sonogram. If abdominal swelling is not as expected (for example, seemingly too large), a scan during the first trimester can easily determine whether more than one baby is present.

The third-trimester ultrasound is done for many of the same reasons that tests are performed earlier in pregnancy. It may also be used to follow up on previous findings, such as a low-lying placenta. Is it still in the same position, or is it now in a more normal spot, higher in the uterus? Another reason may be to check on the volume of amniotic fluid if there was a previous concern about too much or too little. For women who had a previous cesarean section, a third-trimester ultrasound can help rule out an overly aggressive placenta that has invaded too far into the uterine wall. The scan also determines the orientation of the baby. Head pointed down is the optimal position for birth.

Another major reason for a later, non-routine scan is to follow fetal growth, particularly among mothers with preexisting or gestational diabetes, hypertension, or preeclampsia. It may be administered any time there is a suspicion that growth is abnormal (too little or too much).

In relatively rare cases, a concerning finding from a marker blood test or an ultrasound exam may indicate that a woman should have a more detailed ultrasound. Other reasons for this type of scan include being in a high-risk group, such as having had a previous pregnancy in which the fetus had a genetic problem causing a structural abnormality that can be seen with ultrasound imaging.

A more detailed scan can target a particular part of the baby's body or the placenta. Or it may be a stepped-up version of the standard second-trimester scan, essentially taking a more thorough look at the baby. In either case, the test should be performed and interpreted by a highly trained clinician who routinely does this type of advanced imaging. Though you may feel stressed and want quick results, don't hesitate to ask about the experience and qualifications of the person who is administering the scan. In many cases the difference between normal and abnormal is a tiny blip visible only to the highly trained eye. If a problem is found, experts recommend a diagnostic test — the chromosome microarray analysis already described.

OTHER IMAGING METHODS

Magnetic Resonance Imaging (MRI)

MRI is a technology that is particularly adept at imaging soft tissues, such as the brain and spinal cord, rather than bone. MRIs work by putting the human body in a magnetic field. Though studies have not revealed any risks to mother or baby, experts have theoretical concerns, and clinicians have less experience with this technology in pregnancy than with ultrasound. Therefore it isn't routinely used.

Instead, MRI is reserved for special circumstances. For example, it may be used to determine if the placenta has become overly invasive and, if so, how far beyond the normal boundaries the branches of the chorionic villi have spread. But there is an ongoing debate about its accuracy as compared to ultrasonic imaging.

X-Rays and Computed Tomography (CT Scans)

Under what circumstances would a pregnant woman have an X-ray or a CT scan when these methods, which employ ionizing radiation, could harm the baby? The exposure might be inadvertent, occurring because a woman does not know she is pregnant. (This is why women in their reproductive years are questioned closely about whether they could be pregnant before they have this type of imaging, including mammography and dental X-rays.) In the case of an accidental exposure, experts can calculate the chances of a birth defect from the dose of ionizing radiation that was used for the study.

But what if, unfortunately, you sustain an injury during pregnancy, such as a broken bone that requires diagnosis and treatment? When an expectant mother needs to have imaging that uses ionizing radiation because of trauma or a serious medical problem, the benefits of that diagnostic procedure are weighed against the consequences of not getting an accurate diagnosis and the potential worsening of a condition. Experts generally agree that a pregnant woman should go ahead and have the imaging, but it should be done using the lowest dose of ionizing radiation that can produce the image needed to make a diagnosis. If you find yourself in such a situation, work carefully with your clinicians to confirm how best to protect your pregnancy.

Imaging Studies and Breastfeeding

For mothers who choose to breastfeed their baby — what should you do if you discover the need for an imaging study before your child is weaned? The good news is that most imaging technologies — including MRIs, X-rays, and CT scans — are considered safe because they have no impact on breast milk. This is because the agent that is used to produce the image — a magnetic field or ionizing radiation — does not linger. It's there for a fleeting instant, then gone, leaving as its only trace the picture it produced. The exceptions are advanced versions of some tests, which employ substances called contrast agents or dyes in order to enhance an image. Although these usually do not affect lactation, make sure the radiologist or technician is aware that you are breastfeeding and confirm that the contrast agent you receive is safe.

THE NEW TECHNOLOGIES AND YOUR PREGNANCY

Technological advances have changed the nature of prenatal care. It's no longer possible to float blissfully through pregnancy, meeting your baby for the first time at birth. There is a lot more poking and prodding, both physical and mental, than most pregnant women, especially first-time mothers, anticipate. Some expectant women embrace these advances in screening and diagnostic testing because they gain peace of mind from knowing that their pregnancy is progressing normally. Others recoil. But here's the reality: inadequate prenatal care is risky for both a mother and her baby. If you are scared, worried, or turned off by all the new technology, it's important to share those negative feelings with your health care provider rather than flee the medical system. You are in charge! And now you have the information you need to make informed choices about prenatal testing, for your child's well-being and your own.

TAKE CHARGE: RECOMMENDATIONS FOR PRENATAL TESTING

Most prenatal testing is done in the second trimester, though cell-free fetal DNA testing can be done as early as week ten, and ultrasounds may be done early or late in a pregnancy. Here's a quick summary of what most experts on pregnancy suggest:

• *For all pregnant women:* a second-trimester ultrasound

• *For those under age thirty-five:* ultrasound combined with a multiple marker test to screen for fetal chromosomal abnormalities and other potential birth defects, and, if indicated, a follow-up diagnostic test such as amniocentesis or CVS, as well as additional ultrasound scans as recommended

• *For those thirty-five or older:* a first-trimester screening test that analyzes cell-free fetal DNA in maternal blood and, if indicated, a diagnostic test such as amniocentesis or CVS with follow-up ultrasound scans as recommended

4

Protecting Your Pregnancy in the Age of Environmental Chemicals

W omen who are thinking about having a baby or are already pregnant tend to be especially mindful about exposures that they know could be potentially harmful. It's sensible to avoid known risks such as recreational drugs, alcohol, and cigarettes. Science has made the facts clear, and it's not hard to imagine the effects these substances may have on a developing baby.

But when it comes to environmental chemicals (ECs) — manmade substances found in our homes, workplaces, air, food, and water — it's harder to link cause and effect. The risks that ECs may pose are less apparent. If you smoke cigarettes while you're pregnant, the adverse outcomes are well documented. But what if you are exposing yourself and your baby to invisible but toxic chemicals found in plastic food-storage containers . . . the cleaning agents we use in our homes . . . furniture . . . pesticides . . . even personal care products, including the sunscreen that protects your skin from harmful UV rays?

In the twenty-first century, a pregnancy unfolds in a world that is flooded with chemicals. But before you lock yourself away in a protective bubble, rest assured that most environmental chemical exposures we experience as we go about our daily lives are harmless to us and to our children, including a baby who has yet to be born. But some of them indeed pose known risks to mothers and babies while others remain a question mark. Here is where information about the bad actors is key to making informed choices.

The Environmental Protection Agency (EPA) has cataloged about 80,000 manmade ECs, and many of these chemicals find their way into our environment, where they can linger. Current estimates suggest that a pregnant woman is exposed to hundreds (if not thousands) of ECs daily. The CDC has "biomonitoring" data — that is, chemical levels as measured in blood, urine, and other types of samples taken from adults and children — on only a fraction, about 350 of them.

Why is there such a large disparity between the number of ECs we encounter and the number of chemicals that we have data for? The explanation lies in who is studying a particular substance and why. In some instances, the effects of an environmental chemical are investigated in a government or an academic lab setting, with a relatively small research budget that is funded by our tax dollars. Researchers in these labs tend to study a particular chemical when mounting evidence suggests that it has negative effects on human health. The results are made public, so the people who paid for the work through their tax money can benefit from the findings.

But the more involved and expensive studies tend to be done and paid for by the chemical industry itself or a company or private entity that creates or uses the chemical. These multi-billion-dollar businesses are highly motivated to keep their products on the market. Private companies are not legally obligated to release the findings of studies they carry out, and in most cases, the reams of data they produce do not see the light of day. If a commercial product contains an environmental chemical that turns out to sicken consumers, that product sometimes goes away or gets reformulated with a different EC. But making such a change can take years because of the amount of data that needs to be generated to prove that an EC is harmful — work that needs to be done outside the company by academic and government researchers. The slow nature of the vetting process means that companies can put products on the market made with chemicals that they know are risky or have not yet been proven to be safe.

Occasionally, an EC will be banned outright because the scientific data irrefutably shows it to be exceedingly harmful, if not deadly. (One recent example is triclosan, banned by the Food and Drug Administration. Triclosan was used in antibacterial soaps. It was found to be both ineffective *and* linked to hormonal disruptions as well as muscle weakness.) But until they get pulled off the market, you'll find potentially dangerous ECs in a staggering array of everyday items, from cash register receipts to hand lotion to the chair you're sitting in. You can't eliminate them and live in the twenty-first century, but you can take charge to reduce your exposures.

We're about to dive deep into some chemical soup, but here's the good news: even if you are already pregnant, it is *never* too late to start taking action, learning about ECs, determining which ones to avoid, and staying updated about new ones that are making their way into our environment. This information is your best defense against the potentially harmful effects of ECs and the key to protecting yourself and your growing family.

ENVIRONMENTAL CHEMISTRY 101

Even if you wake up on sheets made from 100 percent organically grown natural fibers, you can't make breakfast, get dressed, and prepare for your day without encountering chemical substances — the nonstick skillet you prepared your eggs in; the toothpaste that includes compounds such as cavity-preventing sodium fluoride, dyes, and sweeteners; the makeup or moisturizer you use; the rayon in your shirt . . . and you haven't even gotten to work yet.

These are environmental chemicals, substances that are manmade or extracted by a chemical process. The CDC defines the term as "a chemical compound or chemical element present in air, water, food, soil, dust, or other environmental media such as consumer products."

It's impossible to escape ECs because they are ubiquitous in everyday life, but they are not necessarily bad for us. Many are "inert," meaning that they do not trigger biological activity that might, for example, cause abnormal cell growth or hormonal fluctuations. (The "inactive" ingredients in your toothpaste may be exactly that — they have been included because of their positive impact on texture, taste, or appearance.) On the other hand, some ECs may, in combination with the human body's chemistry, behave in harmful unanticipated ways. They might disrupt normal physiology, interrupting hormone actions and causing health problems.

Some exposures to ECs are more obvious than others, and there are ways to avoid them. If you buy conventionally grown produce, you can assume that it was treated with pesticides, herbicides, fungicides, or chemical fertilizers. You may not know which chemicals were used, but pesticides of some form were probably involved. To reduce your consumption of ECs found in food, buy organic. Or you may be filling up your car with gas and inhaling some of the fumes — among them, benzene, a gasoline additive and toxic chemical that is known to cross the placenta. (Benzene is also used in cleaning solvents, paint strippers, and other products.) If you are pregnant, you can avoid this risk by having someone else pump the gas for you during this particularly sensitive period of your baby's development.

Other toxic exposures are less obvious. Gas-powered vehicles spew air pollutants, which in most cases are invisible. Many of us are oblivious to the potentially harmful flame retardants found in home furnishings, from couches to carpet padding. You may have no way of knowing whether these chemicals are in your home, and if they are, what their potential health effects might be. Some of these substances are so new and poorly studied that we don't really understand what they do to people who encounter them daily. This makes it exponentially harder to fathom their effects on a developing baby and its placenta inside the mother's body.

While there is much we don't yet know about environmental chemicals and their impact on human health and well-being, one important fact has been established. Genes play a large role in whether or not we (and, by extension, our children) are at risk for certain diseases and health issues. But decades ago scientists proved that our whole life story is not written in our genes. The environment — the air we breathe, our water supply, the earth in which we grow our food — plays an equally large role. And today nearly everyone is exposed to ECs.

For example, when lung cancer develops in people who have never smoked, the disease can be traced to a gene mutation — a change in DNA sequence that disrupts normal cellular activity, ultimately resulting in unchecked cell growth that produces a tumor. But in most cases of lung cancer, smoking is the cause, because cigarette smoke is laden with toxic ECs. In other words, different mechanisms — one genetic and the other environmental — can be implicated in the same tragic outcome. An EC, like a gene mutation, is capable of "breaking" a vital part of a cell's machinery. More and more, prominent scientists such as Francis Collins, the current head of the NIH, are calling for studies of disease relationships involving genes *and* the environment and how these two colossal forces interact.

The prevalence of ECs in daily life and the lack of information about their health effects in children and adults as well as pregnant women and their developing babies raise many questions that deserve answers. What is the evidence that these compounds are safe for reproductive health and pregnancy? And why aren't there more regulations governing their use? Most important, how can we take measures to limit exposure to environmental chemicals?

The best antidote to feeling overwhelmed by all these questions and concerns is the same one I recommend for virtually every aspect of a twenty-first-century pregnancy: take charge and become informed.

WHY OBSTETRIC CARE PROVIDERS
OFTEN AVOID DISCUSSING ECs

"Why doesn't my health care provider send me home with a list of ECs to avoid and the products that contain them? There's a 'dirty dozen' list of pesticide-heavy fruits and vegetables to avoid — why isn't there a 'dirty dozen' for the worst ECs too?"

Good questions! Actually, there *are* lists you can consult, and we'll get to them later in this chapter, but the situation is complicated because they aren't comprehensive. You're probably already familiar with some of the known culprits and where they are commonly found. Here are some specific examples of ECs and their sources. Ample data confirms that these substances can be harmful, and governments, including those of the United States and the European Union, have accordingly taken action. Some monitor levels of these ECs, and others ban their use.

* Bisphenol A, or BPA (plastics)
* Flame retardants (furniture, carpet padding)
* Lead (water supply, pre-1978 house paint)
* Mercury (fish)
* Polychlorinated biphenyls, or PCBs (water supply)
* Pesticides and herbicides (conventionally farmed foods)
* Phthalates (plastics)

Of course, this is by no means an exhaustive list, and the routes of exposure are far from comprehensive. A harmful manmade chemical rarely has a single source. These substances creep into all sorts of manufactured products, many of them seemingly benign, and sometimes these chemicals don't stay put. For example, a flame retardant used in manufacturing furniture leaches out over time. When you move a couch, those dust bunnies you find may not be just a harmless nuisance — they are likely laden with a flame retardant. An herbicide like glyphosate (you may know it by its brand name, Roundup) is applied to wheat, so theoretically it could wind up in a loaf of non-organic bread. But it also leaches into the soil and the water supply. Glyphosate, which the World Health Organization has deemed a "probable carcinogen," has shown up in trace amounts in beer and wine — presumably the hops and grapes were sprayed with it.

Another complicating factor is that many ECs aren't exactly household names. It's hard to avoid a harmful substance if information about it is in short supply. For instance, PBDEs (polybrominated diphenyl ethers) are used in the manufacturing of plastics and foam as flame retardants, and research shows that the accumulation of these substances in the human body can cause harm. Studies have offered proof strong enough to earn PBDEs a place in a CDC database of toxic chemicals. This fact provides motivation to avoid bringing PBDEs into our homes in the first place. But you won't find a complete listing of everything that went into the manufacturing of your new upholstered armchair, computer, or big-screen TV — the kind of ingredient labels you find at the grocery store when shopping for food. It's hard to find out where PBDEs lurk.

So you can see why it's all but impossible to get an A-to-Z directory of ECs and their sources from your health care provider. Still, it's important to discuss any possible chemical exposures as part of your medical history during a preconception or prenatal visit. In fact, professional organizations such as the American College of Obstetricians and Gynecologists and the American Society for Reproductive Medicine recommend that clinicians collect that information to enable them to counsel pregnant women and their partners on prevention measures specific to where they live, the jobs they do, and the lifestyle choices they make.

You may be surprised to learn, however, that it's very likely you will sail through pregnancy without a single mention of environmental chemical exposures. There are many reasons for this. One is that your health care provider probably works in an extremely busy clinical environment and has limited time to spend with patients. As a result, the major red flags will be the focus of your visits — a history of pregnancy problems such as miscarriage or chronic health issues such as obesity, hypertension, diabetes, or some combination thereof.

Another reason for this general reluctance to address EC exposures is that reproductive health care professionals may feel uncomfortable discussing the consequences, which are often uncertain. Typically, their clinical training has a strong data analysis component; medical students are taught to read and interpret medical literature and the results of studies by reaching easily understood conclusions reduced to numbers. For example, "folic acid at a dose of about 0.5 mg per day is expected to decrease the risk of a neural-tube defect by an estimated 85 percent in (pregnant) women." There is no mistaking the clear message conveyed by accumulated studies on this topic — take a folic acid supplement!

Unfortunately the data about the reproductive effects of ECs are rarely as clear, except for the grim outcomes that have become world-famous — for example, the brain damage in babies born to Japanese women who consumed mercury-tainted fish in the 1950s (discussed later in this chapter). In that instance, the relationship between cause and effect was clear, backed by data. But where there is a fuzzier link between a specific EC and poor pregnancy outcomes, many prenatal care providers may not feel comfortable giving advice on the subject.

The truth is that studies of the human health effects of chemical substances yield conclusions that *can* be reduced to concrete numbers and estimates of relative risks, data that can then be translated into practical guidelines for all of us. One of my colleagues, Dr. Tracey Woodruff, has been trying to do just that, and she's turned her research into a valuable "Navigation Guide," a practical online resource that your clinician — and you — can use (more on that later in this chapter).

Tracey's background is relevant here: she joined my department after working as a senior scientist and policy adviser at the EPA, where she had become particularly concerned about pregnant women being exposed to ECs in their everyday lives. Eventually she was able to gather biomonitoring data on prenatal exposures from a government survey, which showed that *all* the pregnant women participating in the study had measurable levels of the fifty-two ECs being investigated. These included phthalates, residues from burning coal and gas, flame retardants, pesticides, metals, and a component of cigarette smoke.

Over time she would do additional experiments and research based on blood samples taken from pregnant women or from mothers and their babies, searching for — and finding — measurable levels of ECs in every instance. What Tracey ultimately documented was proof that the placenta acts as a conduit for many ECs rather than a barrier against them.

Before we discuss ways of taking action to reduce your exposure to ECs, it's useful to focus on the reasons for doing so. Let's begin with the simple-to-understand science that shows how ECs work in our bodies, particularly during pregnancy.

HOW ECs CONFUSE THE BODY AND HARM REPRODUCTIVE HEALTH

Endocrine glands are distributed throughout the body. Some, like the thyroid gland, are found in both men and women. Others, like the ovary and the testis, are sex-specific. They all have one thing in common: they release hormones. These molecules travel throughout the circulatory system, delivering powerful signals that regulate basic aspects of our physiology.

We now know that environmental chemicals can interfere with the actions of certain hormones. One of the best-known examples of this is bisphenol A, or BPA, which mimics estrogen, made primarily by the ovaries in women. (Men also have small amounts of estrogen.) BPA is used in the manufacturing of plastics and shows up in everything from food wrap to cash register receipts to cans that contain foods such as soup or tomatoes. Researchers have linked BPA to a variety of health issues, including cancer, dementia, and early puberty in girls. Another example is flame retardants; they can act like thyroid hormone, which has many functions, including regulating metabolism.

The cells that respond to these hormones become confused. They

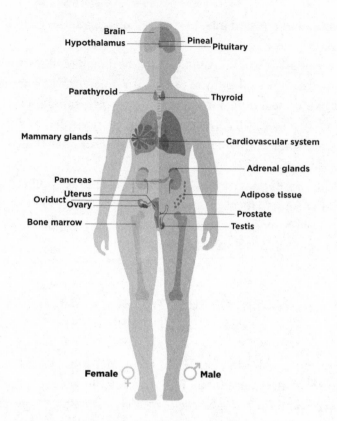

Endocrine glands are located throughout the female body (*left*) and the male body (*right*). They produce powerful hormones, which are released into the blood. Many cells have specialized receptors that bind to these molecules, a lock and key–type mechanism that controls basic aspects of physiology. Some ECs mimic the activities of particular hormones. They have a keylike shape that fits into some of the "locks," but they can't open all parts of the mechanisms. As a result, they disrupt normal physiology and, consequently, health.

don't know whether to listen to the natural version produced by the body or the manmade mimic. It's like hearing two people talk at once, and the messages get garbled. It is easy to imagine how disrupting the actions of sex hormones could have negative effects on reproduction. Likewise, adequate maternal levels of thyroid hormone are required for a baby's normal prenatal growth, including brain development.

ECs and Lower Fertility

My colleague Dr. Patricia Hunt, a professor at Washington State University, by chance linked BPA to reduced fertility when she was studying fundamental mechanisms of reproduction in mice. Pat was conducting research at a lab in Ohio when, mysteriously, she began seeing chromosomally abnormal eggs in the ovaries of normal mice used as controls in her study. These mice also had stopped reproducing — very unusual for animals that are famously fecund. Pat was desperate to find out why. Without fertile mice her research program could not continue. It turned out that they were drinking water from damaged bottles, which were leaching BPA. She found that animals dosed with this chemical had severe defects in their eggs and sperm, and this reduced their fertility.

Previously, Pat had not studied ECs, but after her chance discovery, this work became a large part of her research. Almost exactly twenty years later, working in a new facility in Washington State, she encountered similar problems with fertility in lab animals. In this case, damaged caging materials were leaching chemicals the animals ingested, but this time the culprit was not BPA. Instead, it was one of the replacement bisphenols used as a substitute for BPA in response to consumers' concerns. (These substitutions are discussed later in this chapter.)

Can BPA likewise have a negative impact on human fertility, and also pregnancy? It seems likely. Scientists have found that chemical exposures can be associated with human reproductive outcomes. High levels of BPA were found in the blood of women with a history of recurrent miscarriages. Exposure to BPA during pregnancy has also been connected to aggression and hyperactivity in children, particularly girls.

But of course, it's not just BPA. Ironically, drugs and medications formulated to help people, not harm them, can have unintended negative impacts. For example, nurses who administer certain chemotherapeutic drugs have an increased risk of miscarriages, as do health professionals who are exposed to anesthetic gases used during surgery.

When Babies Are Exposed to ECs

Health conditions related to chemical exposures before birth can manifest years later. For example, farm workers, many of whom are women, have higher levels of pesticides in their bodies than the general population does; there is now strong evidence that children born to mothers exposed to pesticides are more susceptible to childhood leukemia.

Today most people are well aware that eating seafood with high levels of mercury — particularly large fish such as swordfish and tuna — is a health risk and can damage human cognitive function. But we didn't know about the connection between such fish and their disastrous impact on pregnancy until one of the most graphic examples of environmental chemical exposures came to light. In the 1950s, a factory in western Japan released wastewater contaminated with mercury into an ocean bay. About ten thousand people were exposed to high levels of mercury when they ate fish — a staple of the Japanese diet — contaminated with this metal. Many children of women who were pregnant at the time had brain damage with effects that persisted into adulthood.

In the 1960s and '70s, people living in Japan and Taiwan were exposed to very high levels of PCBs — polychlorinated biphenyls — chemicals found in electrical transformers and plastics that have various other industrial uses. They ingested PCBs not through contaminated water, dirt, or air, but through food — rice oil that contained these chemicals, which had leaked from a heating coil.

For pregnant women in these communities, the outcome was particularly grim. The rate of miscarriages increased, and those babies that survived had poorer cognition and lower IQs compared to children not exposed in utero. This type of evidence led the EPA to ban the use of PCBs in 1979. But they are still being released into the environment because the products that contain them — including certain adhesives, oil-based paints, fluorescent light ballasts, electrical equipment, floor finishes, and other items produced before 1979 — are still in use. Furthermore, PCBs are considered extraordinarily "stable," meaning that they don't tend to break down and degrade into smaller and less harmful components. Instead, their molecular structure holds together, and these malingerers "bio-accumulate" up the food chain; eventually humans ingest them. More than forty years after these ECs were banned, the government still biomonitors them.

In the case of mercury and PCBs, the associations between a particular EC and its outcomes were relatively easy to pin down because of the unusually high levels of the toxic substance detected in places where people, including pregnant women, had obvious and sometimes dramatic symptoms. In other instances, these connections are harder to observe and confirm. Sometimes larger numbers of people may be exposed to lower EC levels, possibly leading to subtle biological effects that take longer to manifest but ultimately have a broad impact. Conclusions are harder to draw when people are simultaneously exposed to hundreds or even thousands of chemicals, a situation that may seem unlikely, but isn't. And the outcomes may not be evident for years.

Cigarettes fall into this category. Their smoke is like a complex emission from a high-temperature chemical furnace, containing so many toxic products, such as nicotine and formaldehyde, that it's hard to identify them all. Nevertheless, researchers who follow the effects of maternal cigarette smoking are learning more about the consequences for babies who were exposed before birth. Chapters 1 and 2 describe the links between smoking, low birth weight, and related complications for a newborn, but it is likely that more effects arise when the baby grows into adulthood. Recent data suggest an association with obesity and metabolic syndrome (high blood levels of sugar and harmful lipids, respectively) with the accumulation of fat around the waist — all risk factors for heart disease. The effects of prenatal exposures can last a lifetime.

Still, even though smoke contains a complex mix of chemicals, it is relatively simple to connect the dots — cigarettes are the source, and you know you should avoid their smoke. However, drawing such precise associations between complex chemical exposures and human health gets exponentially harder when the sources are difficult or impossible to identify. This is in fact the case for many of the ECs in our environment.

BEWARE OF THE "REGRETTABLE SUBSTITUTION"

What happens when an environmental chemical receives bad press?

In the case of bisphenol A, as research increasingly pointed to its harmful effects, consumer pressure mounted and it fell out of favor. The FDA banned BPA from baby bottles and sippy cups in 2012. (It is still used in the manufacturing of some products. If a plastic item is labeled with the recycling number 7, which is the designation for polycarbonate, the item likely contains BPA. Today these plastics are recycled into other manufactured goods, such as plastic lumber, but not food containers.)

It took a good long while to call out BPA as troublesome. It isn't a "modern" chemical; it was synthesized in 1881, and its estrogen-like properties were described in the mid-1930s by a scientist who was searching for synthetic versions of the naturally occurring hormone. (He eventually identified diethylstilbestrol, or DES, now infamous for causing vaginal cancer in the daughters of mothers who took this drug for various medical reasons during pregnancy.)

When scientists found compelling evidence that bisphenol A exposures had negative health consequences, chemical companies tried an approach common in the industry: the "regrettable substitution." An old chemical with a long rap sheet is swapped for a new one for which there is very little data related to safety. A familiar pattern begins — levels of the alternative chemical are measured in human samples, safety testing is done in laboratory models, scientists look for harmful effects in the population, and government officials use all of this information to build a case for regulation. This often-repeated

Bisphenol A

$$HO - \bigcirc - \overset{\overset{\displaystyle CH_3}{|}}{\underset{\underset{\displaystyle CH_3}{|}}{C}} - \bigcirc - OH$$

Bisphenol S

$$HO - \bigcirc - \overset{\overset{\displaystyle O}{||}}{\underset{\underset{\displaystyle O}{||}}{S}} - \bigcirc - OH$$

Bisphenol A and bisphenol S have very similar structures and probably share a portion of their biological effects.

cycle burns up time during which the chemical companies continue to reap profits.

The outcome in the case of BPA: products that are now labeled BPA-free often contain this chemical's close cousin, bisphenol S (BPS). Mounting evidence suggests that BPS causes defects in brain development and behavior problems in lab animals. The two compounds have very similar structures, which suggests that they might share a portion of their biological effects — and therefore behave in similar ways when they enter our bodies. The "S" version of this chemical could also have different and even more undesirable effects on human health. It will take time before the amount of data that has been amassed for BPA is available for BPS. And the story of BPA and BPS is not an isolated case. Similar substitutions have been made in many other classes of chemicals, including pesticides, flame retardants, and phthalate plasticizers.

If you see a product, such as plastic wrap or a food container, trumpeting its "BPA-free" credentials, make sure that BPS hasn't been swapped in as BPA's replacement. And as a rule, do not microwave food in plastic containers (the high temperatures draw out the chemicals), and limit your consumption of water and milk sold in plastic containers. Better yet, use glass for food storage. (More tips on avoiding ECs are provided later in this chapter.)

ECs AND PLACENTAL HEALTH

Years ago, I attended a medical conference where speaker after speaker made the same point: environmental chemical exposures are clearly connected to common human diseases. Back then, concern was focused on different toxic substances and their associated health problems: lead and cognitive damage, asbestos and mesothelioma, dioxins and cancer. When a presenter reviewed the evidence linking environmental chemical exposures to low birth weight, all I could think of was the role that the placenta plays in proper growth of the baby. Were these substances somehow circumventing the placenta's job as nourisher and gatekeeper?

The answer, as Tracey Woodruff's research shows, is yes. The graph opposite is from one of her studies, in which she measured levels of environmental chemicals in the blood of mothers and their babies. Each bar illustrates the data for one mother (graph on the left) and her baby (graph on the right). Remarkably — though not surprisingly — the baby's bar is a reflection of the mother's. The participants had measurable blood levels of up to forty of the environmental chemicals researchers were looking for, including metals, flame retardants, and pesticides.

These findings raise many questions about how chemicals affect the developing baby and its placenta. Take, for instance, PBDEs, the flame retardants mentioned earlier. They are of particular concern in my home state, California, because for many years, state law held furniture manufacturers to a fire-safety standard higher than those in other parts of the country. The law, meant to protect consumers, had an unintended and unfortunate result. To meet the fire-safety requirements, manufacturers added more PBDEs, a potentially toxic class of chemicals, to their products. In practical terms, this

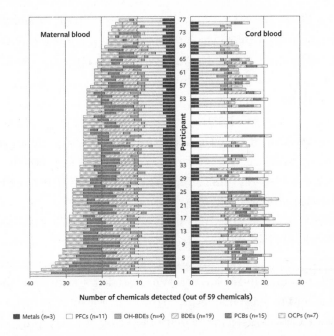

Every sample that was obtained at the time of birth from each participant, either the mother (maternal blood) or her baby (umbilical cord blood), had detectable levels of many environmental chemicals. The different shades depict the chemical classes that were detected, for example, metals, perfluorochemicals (PFCs), flame retardants (OH-BDEs and BDEs), PCBs, and organochlorine pesticides (OCPs). The blank areas in the bar graph to the right denote missing samples, not an absence of chemicals. (Modified from Morello-Frosch et al., 2016.)

meant that the average couch contained two to three pounds of flame retardants — and keep in mind that this furniture was not sold exclusively in California.

This extreme safety standard was overturned in 2014, but PBDEs are classified as "persistent" — meaning these chemicals do not break down easily. It's expected that PBDEs will remain at high levels in the environment for years to come, and it's easy to understand why. Few people can afford to replace their furniture simply because it has an invisible payload of flame retardants.

Dr. Joshua Robinson, my colleague at the University of California

San Francisco, discovered that commonly used flame retardants can disrupt some of the networks that control how brain neurons develop. Other studies link prenatal exposures to flame retardants to possibly related effects in school-age children: poor attention span, compromised fine-motor coordination, and cognition challenges. In fact, prenatal exposures to PBDEs can lower a child's IQ almost as much as exposures to lead, a well-studied connection. My group found that these compounds disturb the inner workings of placental cells, which supports findings that link prenatal exposures to flame retardants and reduced fetal weight.

Phthalates are another example of pervasive ECs that can pass from mother to baby via the placenta. As a group, phthalates appear in a broad range of everyday products, from vinyl flooring and food containers to inks and cosmetics. When used as "plasticizers" they can soften otherwise hard materials, so they become more pliant. Phthalate plasticizers are used in the manufacturing of PVC piping, shower curtains, garden hoses, rain boots, window frames, auto interiors, children's toys, and more. In short, they are hard to avoid.

Previous studies have raised plenty of red flags. One finding suggests that prenatal exposures to phthalates can shorten the length of pregnancy. Other research shows that effects of phthalates on children differ by sex, indicating that these ECs may be disrupting development in utero. In boys, prenatal phthalate exposures adversely affect male reproductive development and may reduce "masculine play" (that is, choice of toys and activities). In girls, consequences include impaired neurodevelopment as well as a link to early puberty.

Here's an example of how hard it is to predict where and how we are exposed to ECs like phthalates: When Tracey Woodruff first embarked on studying blood samples drawn from pregnant women, the initial batch collected for testing had to be discarded; ironically, the plasticized tubing through which the blood was drawn leached high levels of phthalates that contaminated the samples. Once the blood collection protocol was redesigned, Tracey and her group showed that the newly collected samples contained many phthalate metabolites, products our bodies make by breaking down these chemicals. Ultimately, the results of her work are consistent with the conclusion that developing babies are exposed to ECs throughout pregnancy, starting at conception.

Why are we and other scientists doing so much research on ECs? These aren't academic exercises driven purely by scientific curiosity. Government regulators and policy makers use the results of studies like ours to build a case for limiting or even banning the use of particular environmental chemicals. These decisions must be evidence-based, and not all types of data are given equal weight. In general, results obtained by modeling the health effects of exposures using human cells carry more weight than animal studies or experiments on simpler organisms. Because our lab focuses on human development, we're perfectly positioned to do this type of research.

Tracey's measurements of ECs in pregnant women and their babies are examples of studies that can — and do — impact government actions. PBDE flame retardants are now being phased out nationally, and California banned them years ago. Furthermore, since Tracey and her group have been collecting data, the levels found in pregnant women have decreased by 60 percent. This shows that when the government is willing to act, its people reap the benefits.

For decades, due to lack of oversight, loose laws, or outright deregulation of these harmful substances, policy has prioritized the economic health of chemical companies and other businesses responsible for the release of ECs over the physical health of citizens. Through our research and that of our colleagues, we hope to shift the momentum in the opposite direction, making the health of pregnant women and their families — and all children and adults — the highest priority.

WHAT YOU CAN DO ABOUT ECs

The threat of ubiquitous ECs may seem overwhelming, but you do have the power to protect yourself and your growing child from potential risks. Here are some simple, everyday steps you can take, starting now.

Assess Your Risk of EC Exposures

If you haven't had an opportunity to discuss ECs with your health care provider (whether you are pregnant yet or not), be proactive and do a risk assessment using the questions below before your next appointment. This is a good start to determine your potential risks, will make you more aware of ECs in your everyday life, and will enable you to take charge and reduce exposures. Your answers will also prepare you for discussions with your clinician.

* What is your job, and does it expose you to metals, solvents, chemicals, radiation, or fumes?
* Do you live in a house that was built before 1978? (The federal government banned the use of lead-containing paint that year.)
* Does your house have lead in its pipes?
* Are you planning to remodel your house? If so, what materials will be used?
* Do you use weed killers or pesticides such as bug killers or rat poison?
* What chemicals, if any, do you use for cleaning?
* Do you take herbal remedies? If so, which ones?
* Do you use fragrant personal care products? Which chemicals are listed as part of their ingredients?

Of course, some exposures are more serious than others. Working in an industry where you come into daily contact with ECs is drastically different from occasionally using an ant killer. But either way, you can protect yourself and your baby from ECs by taking action. This may involve the simple choice to avoid certain household products and personal care items. When you reduce EC exposures to safeguard you and your child, you're also helping to reduce the risks they pose to everyone around you by cleaning up the air, protecting our shared food supply, and keeping our water safe. Or you might want to go further, seeking broader protection for everyone by supporting federal and state laws that guarantee safety in the workplace, for example.

Solve the Plastic Problem

Plastics are polluting the planet — and some ECs found in plastics, such as BPA, BPS, and phthalates, have the potential to pollute our bodies as well. But realistically, you can't avoid plastics entirely. You can, however, minimize your everyday exposure to these ECs, especially during pregnancy. Here are a few practical suggestions.

* *Store your food in glass, not plastic.* If you tend to refrigerate your leftovers in plastic containers, stop. Even when labeled as BPA-free, these containers may contain a "regrettable substitution," such as BPS. Many people are tempted to reuse takeout containers or the packing from store-bought foods, but put them into the recycling bin instead. Invest in inexpensive glass containers, and if you can, avoid using clear food wrap. It may be made of a plastic film, polyvinyl chlorine (PVC), that can leach another plastic, di(2-ethylhexyl) adipate, or DEHA, into your food, which may have hormone-like activity.
* *Don't microwave food or beverages in plastic.* Doing so adds plasticizers to your fare. Instead, microwave food in tempered glass, and remove any plastic lids. If you cover the container with plastic wrap to prevent splatter, don't let the film come in contact with your food during heating. Try using a tempered glass plate as a lid.
* *Ditch the old plastic water bottle.* If you carry a plastic water bottle around all day for hydration, especially a bottle that's showing signs of age, you may be drinking ECs. Visible wear-and-tear indicates that the material is disintegrating and likely leaching

chemicals into your water. Glass is widely considered the safest option for reducing EC exposures. But if you are constantly on the go and breakage is a concern, consider switching to stainless steel, which is nonreactive, meaning it won't affect the contents of the bottle. Aluminum water bottles are popular, but they must be lined to prevent metals from leaching into the liquid they contain. The linings — like those in aluminum cans — often contain BPA. Single-use plastic water bottles won't hurt you (unless you make a habit of reusing them), but given how they're contributing to plastic pollution, now's your chance to leave a little less plastic in the world for the next generation by forgoing their use. Consider containers that are labeled "bio-based" or "greenware," which indicates that they're made from biological sources such as corn and contain reduced amounts of ECs, or none at all.

* *Avoid these numbers — 3, 6, and 7.* The American Academy of Pediatrics suggests avoiding plastics labeled for recycling with the numbers 3, 6, and 7. While this advice is meant for parents who want to safeguard their children's health, it's applicable during pregnancy as well. Plastic number 3 may contain phthalates and/or polyvinyl chloride (PVC), which is found in plastic wrap, shower curtains, toys, and other items. Plastic number 6 indicates the presence of polystyrene, a component of the Styrofoam used in takeout food containers. It's been linked to central nervous system damage as well as some cancers. Plastic number 7 includes hard materials such as polycarbonate. It's found in reusable water bottles and cups, often together with BPA.

* *Don't save those receipts.* The thermal paper used for most receipts is coated with BPA or BPS. Do you really need a record of every single purchase or ATM withdrawal? You may be in the habit of politely accepting them, but unless there is a reason, decline receipts. Why hoard a source of EC exposures in your pocket?

* *Stay informed.* In general, aim to use fewer plastic products, and choose wisely when you do. Organizations such as the Environmental Working Group (EWG) have good suggestions on how to do this. See their consumer guides at www.ewg.org.

Breathe Cleaner Air

It's true that during pregnancy, your sense of smell may change in odd ways and become more acute. If you are bowled over by strong odors from places like a just-cleaned bathroom, a nail salon, or an idling car, you're probably inhaling ECs. But if you use common sense you can still breathe easy, and you'll be doing your part to support a healthier environment. Here are a few suggestions:

* *Don't inhale (and do call in the reinforcements).* The easiest solution is to avoid the source of the smell or not to linger in its vicinity. Especially if you're pregnant, get someone else to pump the gas, paint the hallway, or scrub the toilet bowl.
* *Choose nontoxic cleaning products.* Maybe you can't avoid every housecleaning chore, but you can choose cleaning products that are easier on you and the environment. Vinegar, soap, and baking soda are good alternatives to premixed commercial cleaners. Visit websites such as Green Seal (www.greenseal.org) to find green products and services.
* *Banish the dust bunnies.* Dust particles can concentrate ECs. (Consider the dust that accumulates beneath a piece of furniture treated with flame retardants.) Don't let them build up in your home. Wipe down or mop up surfaces where dust gathers. Use a vacuum cleaner with a HEPA filter to trap dust and dirt.
* *Get rid of the cigarette smoke.* You already know that smoking and pregnancy don't mix. The ECs in tobacco smoke, such as nicotine and cotinine, are particularly harmful. Make your home a smoke-free environment.

* *Put out the fires.* Limit exposures to smoke from wood-burning stoves and fireplaces. Wood smoke contains fine particulate matter that holds toxic ECs, such as benzene and formaldehyde. It's easy to breathe in this microscopic matter if you're near the smoke. It's just as easy to avoid it.
* *Wait till the air clears.* If you live near likely sources of air pollution (busy roadways or industrial sites), avoid outdoor exercise on days when pollutants are high. Most cities put out air-quality alerts; you can also check the weather app on your smart phone.
* *Choose clothes carefully.* Dry cleaning releases toxic chemicals into the air and water — it's best to purchase clothes that don't require it. That said, the emissions from clothes dryers powered by gas aren't healthy for the planet either. Though it's more work to hang clothing to dry outside, the sun is the ultimate earth-friendly option!

Do a Domestic Detox

We know from studies on animals and humans alike that the so-called nesting instinct during pregnancy is real. Birds feather their nests, and pregnant women get organized and clean house. But there is more to making a healthy home than choosing green cleaning supplies. The tips that follow are not comprehensive — you'll find entire books on topics such as "eating clean" or "living green." This list is simply meant to jumpstart your thinking about ECs in your home, especially when you're expecting.

What's in the Kitchen?

* *Go organic (or close to it).* The best way to avoid ECs in your food supply is to go organic, which is becoming easier (and a bit cheaper) as consumer demand increases. But not every item

needs to be organic. Download an app such as the Pesticide Action Network's "What's on My Food?" or the Environmental Working Group's shopping guides for advice on which organic foods you should prioritize and which conventionally grown items are relatively safe. (See Resources.)

* *Choose BPA-free cans.* Buying canned foods is not taboo. Just remember that aluminum cans are made with a liner to keep food from coming into contact with the metal, and often this liner is made of BPA. Some brands of canned food do not have BPA liners and will be labeled as such — a better choice. It's even healthier to select "canned" foods that come in glass jars.

* *If you can't "eat clean," then "eat lean."* If you eat animal products, limit your consumption of animal fat, as that's where the toxic substances accumulate (from pesticide-sprayed grain or added hormones, for instance). Consider switching to organic dairy when you're pregnant and breastfeeding. If you can afford the extra cost, it's good to make this change permanently.

* *Keep mercury out of your diet.* Fish is packed with healthy omega-3 fats, but you should avoid those that are at the top of the ocean's food chain, such as swordfish, shark, king mackerel, tilefish, and albacore tuna. Generally, the older and bigger the fish, the higher the concentration of mercury. For more information on choosing safe seafood and fish, see the Monterey Bay Aquarium Seafood Watch website (www.seafoodwatch.org).

Are Those Home Improvements Really "Improvements"?

* *Delay the remodel.* If possible, put off home improvements and repairs while you're pregnant, especially if you're living onsite. Activities such as painting, sanding, and varnishing unleash ECs. Lead paint, used in older homes, was removed from the market in the late 1970s, but it still can be found on unrefurbished surfaces or under newer wall coverings. Sanding aerosolizes this toxic substance.

* *Renovate the right way.* Once you're ready to remodel, choose safe home-improvement materials. Some products — such as pressure-treated lumber, which has an arsenic-based preservative — have been found to have high levels of toxic ECs. Visit the Environmental Protection Agency's website (www.epa.gov), and search for their "Renovate Right" printable resources.
* *Furniture shopping?* If you are buying new furniture, select flame-retardant-free products. (See Resources for a link to shopping guides developed by the Center for Environmental Health.) If you have older furniture, consider replacing the foam in the cushions with new flame-retardant-free material.

How Does Your Garden Grow?

* *Pull rather than poison.* Don't use toxic products for killing weeds and bugs or fertilizing your lawn. Not only are you subjecting yourself to ECs but you're also putting them into the soil and water supply. Whenever possible, tugging weeds out by hand is a better option. You can find "green" lawn and garden products and services in most communities.
* *Adopt an "integrated pest-management strategy."* There are ways of dealing with troublemakers (from weeds to bugs to animals) without pesticides or other chemicals. The University of California offers a useful website describing this environmentally friendly approach. (See Resources.) Whether you want to grow flowers or food, an online search will yield a bounty of websites on organic gardening how-to's.

DOES "NATURAL" ALWAYS MEAN "SAFE"?

Traditional tried-and-true remedies like chamomile tea for an upset stomach are touted as all-natural, centuries-old alternatives to conventional medications. But when you're pregnant, you need to make sure that your great-aunt's all-purpose home remedy is EC-free and safe for you and your baby.

Some folk remedies for common ailments have in fact been associated with lead poisoning. They include azarcon or greta (taken for upset stomach or diarrhea), bo ying (taken for flu and colds), and pay-loo-ah (taken for fever or rash). If you have questions about the safety of natural remedies, speak with your clinician. You can also visit mothertobaby.org to learn more about the safety of various alternative medical treatments during pregnancy and breastfeeding (see Resources). However, it's important to note that the manufacturers of herbal remedies don't have to receive approval from the Food and Drug Administration before they put these products on the market. These remedies are classified as food supplements rather than pharmaceuticals. Therefore any dangers that they pose become apparent only after the ill effects become evident.

ECs AND PERSONAL CARE PRODUCTS

Other than the dyes they contain, we can discover very little information about the composition of soaps, lotions, cosmetics, perfumes, and the like. Just try looking for a list of ingredients for your foundation, eyeliner, or lipstick — in most cases, there is none. In a very real sense, we are using beauty potions — magical mixtures that we hope will have the desired results. But they might also produce unanticipated and undesirable side effects because of their EC content. The long-term subtle effects of repeated exposure to the ECs in these items, some of which are endocrine disrupters, are unknown.

Far more women than men use beauty products and work in the salon industry. Some experts think that this is why the National Health and Nutrition Examination Survey — US government–sponsored studies that include monitoring of environmental chemical exposures — found that women were three to four times more likely than men to have in their systems high levels of parabens, which mimic naturally occurring estrogen. Parabens are chemicals added as antimicrobial preservatives and fragrances to personal care products such as lotions and hair relaxants. (They are also present in food, beverages, and pharmaceuticals.)

Research found that pregnant women who had recently applied skin lotion had paraben levels that were about twice as high as those who had not. In the same population, the use of certain personal care products, such as nail polish, eye makeup, and fragrances, positively correlates with higher levels of phthalate plasticizers. In fact, there is a dose-response relationship between the number of personal care products a pregnant woman uses "and the concentration of phthalate plasticizers in her urine," strong evidence that these products contain this class of ECs.

How can women minimize exposures to environmental chemicals in personal care products? Unless you're ready to give them all up, the alternative is to become a smart consumer. The Environmental Working Group's Skin Deep website (see Resources) gives detailed information about nontoxic products, including cosmetics, skin care products such as moisturizers and sunscreens, hair and nail treatments, toothpastes, soaps, fragrances, and many others. As more and more consumers express concern about ECs in their personal care products, the market is responding with new nontoxic versions. Referring to online resources such as Skin Deep is the best way to keep up with this changing landscape.

HOW TO PREVENT CHEMICAL EXPOSURES AT WORK

If you work in an environment where chemicals are regularly used, such as a salon or spa, a medical lab or hospital, a construction site, or an industrial setting, you may have good reason to be concerned about everyday EC exposures. Keep in mind that a safe workplace is a basic right, whether or not you are pregnant. Here are some helpful guidelines that may be especially useful if you want to approach your employer with questions or concerns.

• Know the laws and regulations. You're entitled to information on the chemicals you may be exposed to. The Occupational Safety and Health Act (OSHA), which covers federal government and private-sector employees, gives you many rights regarding chemical exposures.

• Consider that there are numerous ways to be exposed. They include skin contact, inhalation of fumes or dust, and eating, drinking, or smoking without washing your hands.

• Find out about available accommodations your employer can make to meet your needs while you are pregnant and nursing. The National Labor Relations Act protects both unionized and nonunionized employees from discrimination when two or more individuals act together to request, for example, accommodations during pregnancy.

• If you meet resistance or discrimination related to a request for accommodations, the law is on your side. The Pregnancy Discrimination Act, a federal law, prevents discrimination based on pregnancy.

• Secondhand exposure to chemicals is possible. If a family member works with chemicals, avoid coming into contact with them or their clothing until they have showered, if necessary, and changed.

• Take advantage of the Hazard Communication Standard, a federal rule that mandates how employers tell workers about the chemicals they are exposed to and what the dangers might be.

• At work, pay attention to notices that indicate a danger of chemical exposure and other risks. Read the text and look at the symbols — which are sometimes not easy to decipher. The

surprisingly nonintuitive diagram shown above is the signage meant to indicate "health hazard."

• Ask for the (Material) Safety Data Sheets for products you work with. They contain valuable information about health effects.

• Follow the guidelines provided for chemical use, including recommendations for ventilation, spills, and the wearing of protective equipment.

• Talk to your company's health and safety specialist about your exposure to chemicals. These experts can advise you on protective measures.

• Discuss your chemical exposures with your health care provider, who may be able to offer additional advice and sources of information.

MORE RESOURCES AT YOUR FINGERTIPS

In addition to the consumer-friendly information you'll find through the Environmental Working Group and the Program on Reproductive

Health and the Environment, here are some additional online resources that focus on EC reduction:

* *The Center for Environmental Health,* an organization dedicated to protecting families from chemical exposures, offers tip sheets on topics such as avoiding flame retardants in household products. (https://www.ceh.org)
* *The Collaborative on Health and the Environment (CHE)* has launched a consumer-friendly website that offers a wealth of practical information for families. The CHE's mission is to provide evidence-based science on topics such as ECs, offering a range of hands-on solutions. (https://www.becausehealth.org)
* *The Green Science Policy Institute* translates the science of chemicals into practical information that consumers can use as they make buying decisions on products ranging from furniture to personal care items to baby products. (http://greensciencepolicy.org)

The CDC's National Biomonitoring Program lists substances that all of us — pregnant or not — should be aware of. (See Resources.) If an EC (or metal) is on the list, it means that evidence exists to indicate that it may be harmful to human health or have toxic effects. The inventories are comprehensive and include volatile organic compounds (such as benzene), chemicals used to make plastics (such as BPA), and other everyday synthetic materials present in our environment: flame retardants; pesticides, insecticides, fungicides, and herbicides; toxic metals such as mercury, lead, and cadmium; and much more.

As you look through the various listings and tables, you may find them dense or overly technical. They were, after all, designed with public health officials and other medical professionals in mind, but don't be deterred. With a little patience and careful attention, you'll be able to delve into them. And you should — particularly if you have a question about a specific substance. For each chemical listed, you'll find a useful link to a fact sheet, which includes easy-to-understand information on where the substance is found, how people are usually exposed to it (product usage, air, water, and so on), and the possible health repercussions of these exposures. It's an eye-opening introductory education to the wide array of ECs we live with.

TAKE CHARGE: GO TO THE EXPERTS ONLINE AND IN PERSON

As you become more knowledgeable about ECs, you'll be in a better position to ask your clinician focused questions about them. If your health care provider still doesn't offer specific advice, or if you feel you're not getting enough information, share this resource: The Program on Reproductive Health and the Environment (PRHE), which is based at UCSF. (See Resources.)

PRHE's website features a navigation guide, created by Tracey Woodruff and her colleagues, whose work has been discussed in this chapter. The website is designed for OB/GYNs, pediatricians, and any and all health care professionals who care for reproductive-age women and men, as well as children. The guide reviews the latest literature and bottom-line summaries about EC effects for clinicians, all intended to make them more comfortable about making recommendations to patients. And you too can use the PRHE website. You might find the family-friendly General Resources section especially helpful; it offers a wealth of information on reducing your EC risks at work and at home.

If you have special concerns about your child's pre- or postnatal EC exposures, highly qualified pediatric experts are available to answer your questions and help you make an action plan. You can find them through a national network, the Pediatric Environmental Health Specialty Units. (See Resources.) Pediatricians and related specialists have been dealing for decades with EC exposures in children — for example, high levels of lead in the blood. But generally speaking, most health care providers don't receive extensive training in environmental health issues. Through these units, you, your pediatrician, and your family's other health professionals (including your prenatal care provider) can tap into a network of specialists that serve every region of the country.

It's never too late to reduce your exposure to ECs or to take steps that will protect your family. And by setting an example through environmentally responsible choices, your influence could extend far beyond your own household. When you protect your child, you're protecting the planet too.

5

When Pregnancy Gets Complicated

T he odds of having a successful pregnancy and delivering a healthy baby are overwhelmingly in your favor. This very reassuring fact means that most women and their partners will never *need* to read this chapter. That's right — you may choose to skip it entirely.

But then again there are good reasons to read on. You may be surprised to learn that despite the access to good prenatal care that most women in the United States enjoy, the rates of some common complications are on the rise. If you acquaint yourself with the risk factors, you may be able to increase your odds of having a healthy baby. We'll also cover the warning signs that may indicate major problems in a pregnancy, information that is good to know. And even if you don't anticipate problems, the more you know, the better prepared you'll be to take charge and deal with any challenges — great or small — you may encounter.

Of course, many expectant mothers will experience some relatively minor complaints along the way — garden-variety morning sickness, fatigue, or back pain, for instance. Admittedly, these common symptoms are not minor if they have a major impact on your day-to-day routine, but they are not life-threatening to you or your baby. However, there is no denying that occasionally, some pregnancies unfold with more serious complications. This chapter explains what they are, including their causes, possible preventive measures, and treatment strategies.

If you are facing a more complicated pregnancy than you'd anticipated, maternal-fetal medicine (MFM) doctors, who specialize in treating women with high-risk pregnancies, will be there to help you. These obstetricians have specialized expertise in recognizing the signs of the conditions described in this chapter, and they will know the latest treatment strategies. Having a highly trained professional on your side is a great way to mitigate risks and allay concerns.

Problematic signs can appear later in pregnancy, when initial discomforts (the nausea and fatigue of the first trimester) are behind you. Once a woman reaches the midpoint of pregnancy, she is generally feeling better. The baby is growing, and ultrasound images begin to resemble a child rather than a blurry picture of some alien creature. If all has been going well, it's particularly shocking to be told, at a routine second- or third-trimester checkup, that signs of a pregnancy complication may be appearing.

The most common problems — preterm labor, preeclampsia, and gestational diabetes — are the focus of this chapter.

PRETERM BIRTH

Preterm, or premature, births are defined as those that occur before thirty-seven weeks of pregnancy. The timing mechanism that is normally set for delivery at term (from early term at thirty-seven to late term at forty-one weeks and six days) goes haywire, and the process begins too soon. Definitive signs include premature rupture of the fetal membranes that surround the baby ("water breaking"), which may or may not be followed by the initiation of labor, leading to cervical shortening and dilation. Or labor may start, and cervical changes may occur without membrane rupture. If these processes can't be interrupted, birth will happen before the baby is fully mature and ready to cope with life outside the womb.

Spontaneous preterm labor, the prelude to preterm birth, is not

a single disease with a well-established cause and a set of symptoms. About 30 percent of preterm births are unexplained. While
we don't know why they happen, we do know that the greatest risk
factor for having a preterm birth is having had a previous preterm
birth. Another 20 percent of preterm births are "iatrogenic." *Iat-
ros* originates from the Greek word for "doctor," and this medical
term literally means that iatrogenic births are caused by a doctor (or
medical treatment) through either induction of labor or a cesarean
section. The rationale for going ahead with one of these procedures
could be that the mother has a serious pregnancy-related health condition or signs indicate that the baby has stopped growing or is in
distress.

The reasons for the remainder — about half of all preterm births
— can be explained. In other words, about 50 percent of the time the
cause is understood. Unfortunately, the explanation is often not clear
until after the fact, something that scientists who focus on pregnancy
are trying to rectify, which would enable better prediction of this complication. Let's look more closely at the known causes.

Infections

The most well understood cause of preterm labor is infection of the
amniotic cavity, the fluid-filled space that surrounds the baby (see
page 46). In about a third of all preterm births, the abnormal growth
of bacteria or viruses in this mostly germ-free chamber is the instiga-
tor. Where do they come from? Frequently, they are the same kinds
of infectious agents typically found in the vagina, which suggests that
something has gone awry in the peaceful cohabitation between the
placenta and fetal membranes on the one hand, and the mother's
normal microbiome on the other.

How does an infection in the amniotic cavity trigger a preterm
delivery? It appears that the response of the mother's immune system
to infection involves some of the mechanisms that normally initiate labor. The muscular layer of the uterus seems to get confused,

mistaking the abnormal signals emanating from the infection for signals indicating that the pregnancy has come to term. Therefore the uterus starts contracting.

On the surface, this problem seems to point to a biological design flaw. But viewed through the lens of evolution, there may have been advantages to the survival of our species. Preterm labor likely benefited our ancestors: the mother was able to expel an infected pregnancy, preventing damage to her uterus and increasing the chances of successfully reproducing in the future. And some advantages persist. Preterm birth allows a baby to escape a potentially fatal infected environment, increasing the chances of its survival.

Uterine Issues

As we discussed in chapter 2, the uterine lining acts as a sort of bandage to stanch the flow of blood from vessels that the trophoblast cells of the placenta tap into. If problems with the uterine lining inhibit this clotting process, blood products are released, and some of them may stimulate labor.

Another theory involves the aging of the uterus. All of our organs age, and the uterus is no exception. The related functional decline can impede the elaborate process that transforms the uterine lining into the decidua, a nurturing environment for an implanting embryo. Impaired decidualization is another risk factor for preterm labor (and, as discussed later in this chapter, for preeclampsia).

Uterine stretching, which occurs when a mother is carrying more than one baby, is another factor that sometimes leads to preterm birth, but researchers still don't know why. In a singleton pregnancy, the uterus swells far beyond its normal size. Somehow the organ has evolved to tolerate this enormous expansion (discussed in chapter 2), but there are limits. Mothers carrying twins or other multiples have a greater probability of delivering before thirty-seven weeks. In lab studies with animals, an increase in uterine volume leads to preterm delivery.

Hormone Levels

There is significant evidence that the hormone progesterone plays a role in preterm labor. During a normal menstrual cycle, a woman's progesterone levels rise during the latter half, after ovulation. As a result, the uterine lining thickens and undergoes molecular changes that are critical to establishing pregnancy. If an embryo doesn't implant, progesterone levels fall and the uterine lining sloughs off. However, if an embryo does implant, progesterone levels stay high, which sustains the uterine lining (now being transformed into the decidua). Progesterone also quiets the uterine muscles, making it harder for contractions to begin.

Because of the many beneficial effects of this hormone in the maintenance of pregnancy, a decline in progesterone levels may be an important risk factor for preterm labor and birth. Fortunately, new and effective strategies for preventing preterm labor (discussed later in this chapter) are taking advantage of this fact.

Other Factors

Unmanaged stress plays a role in every person's health, and its dangers are well documented. It can lead to serious medical conditions such as heart disease, and it can weaken our immune responses. Not surprisingly, stress during pregnancy is particularly troublesome.

There is evidence that social stress, in particular, is linked to preterm birth. Scientists who have examined the incidence of preterm birth in relationship to sociological factors have consistently uncovered a strong association between preterm birth and certain life circumstances, such as young maternal age, lack of education, poverty, and being unmarried.

Race is also a significant contributor. In a comparison of women closely matched in the areas of social stress, socioeconomic status, and prenatal care, those who were Black had about twice as many preterm births as those who were white. The reasons, which are under intense investigation, are as yet unknown.

Your risk of having a preterm delivery may also depend on where you live in the United States. Wide discrepancies exist from state to state and in different urban areas. According to the March of Dimes, in 2018, Louisiana, Mississippi, Alabama, and West Virginia had the highest rates of preterm birth in the country, between 12 and 13.5 percent. Oregon and Washington had the lowest, at around 8 percent. Among major cities, Cleveland, Detroit, and Memphis were out in front at over 14 percent, with Baltimore and Baton Rouge following close behind. The cities with the lowest rates of preterm birth included Irvine, Seattle, Portland, and Vancouver, which were 8 percent and below. No doubt race and the effects of social stress contribute to the regional and local disparities in preterm birth; access to health care also figures in.

The statistics on preterm birth are attention grabbing. In recent years, rates have been rising across the United States, hovering now at around 10 percent of all deliveries. (The March of Dimes compiles this information, and updated numbers are posted on its website every year. See Resources.) You would think that in a resource-rich country like ours, the rates would be relatively low, but that isn't the case.

When we put American numbers into a world context, the results are even more surprising. The country of Belarus has the lowest rate, at around 4 percent; the African nation of Malawi has the highest, at 18 percent. Thailand and Turkey report fewer preterm births than the United States, while the Southeast Asian countries of Timor-Leste and Brunei are only slightly higher on the list. It's sobering that our country lies somewhere in the middle of the pack, alongside countries with far fewer resources.

Clearly, there is much to be learned about the causes of preterm labor and birth, especially as there is no simple fix for this pregnancy complication once it arises — in part because every case is an individual odyssey. I learned this firsthand when I was pregnant for the first time — narrowly avoiding a preterm birth after six uneventful months.

There are, however, some factors that you can modify, including taking charge of your own health before you become pregnant, as well as developments on the prevention front, which we'll look at next.

RISK FACTORS FOR PRETERM BIRTH

The lists below can be invaluable to any woman who wants to understand her risk factors for preterm birth. They were compiled by researchers for use by health care professionals. Some factors are difficult or impossible to change, and others are potentially modifiable. Knowing the difference can help you take charge and reduce your risk.

Not Modifiable or Hard to Change

- Prior preterm birth

- Race (prevalence is higher among Black women)

- Age less than eighteen or greater than forty

- Poor nutrition (prior to the onset of preterm labor)

- Low pre-pregnancy weight

- Low socioeconomic status

- No prenatal care (prior to the onset of preterm labor)

- Cervical injury or anomaly

- Uterine anomaly

- Excessive uterine activity (wavelike contractions)

- Premature cervical dilation or effacement

- Carrying more than one baby

- Vaginal bleeding

Potentially Modifiable
- Cigarette smoking

- Illicit drug use

- Anemia

- Infections of the urinary tract or the lower genital tract

- Gum disease

- Strenuous work or work environment

- High personal stress

- Being overweight or obese

It's also important to know the warning signs of preterm labor. The following list is compiled from two sources, the March of Dimes and the CDC. If you experience any of the following symptoms, you should contact your obstetric care provider immediately.

- You have contractions every ten minutes or more often. The abdomen tightens like a fist, and the contractions may or may not be painful.

- There's a change in vaginal discharge: a significant increase in the amount of discharge, leaking fluid, or bleeding from the vagina.

- You feel pressure in your pelvis or lower belly, as if your baby is pushing down.

- You have a constant low, dull backache.

- Belly cramps occur, with or without diarrhea.

- You have cramps that feel like those of a menstrual period.

- Your water breaks.

Preventing Preterm Birth: The Good News

For decades, the medical field made very little progress in preventing preterm birth. Most of the available therapies proved ineffective and were abandoned.

At one time, progesterone therapy was used indiscriminately for women who had a problem carrying a baby to term. Studies showed that treating all patients in this way was not beneficial. But then researchers revisited this approach, exploring whether progesterone might be effective if given in a more targeted way to specific women. The answer was yes — it can be effectively used for prevention in groups having the highest risk.

In 2011, the FDA approved the use of systemic progesterone therapy (delivered by injection) for preventing preterm birth in women who had a previous spontaneous delivery before thirty-seven weeks of pregnancy. Studies showed that this treatment reduced preterm labor by about 20 percent for this high-risk group.

Another group that benefited included women with a short cervix, a condition that can be identified by transvaginal ultrasound from the fourth to the sixth month of pregnancy. Vaginal administration of progesterone before the occurrence of any signs of preterm labor reduced preterm birth by 45 percent.

Previously, cervical cerclage was a common treatment for pregnant women who were at high risk of having a preterm delivery, including those with a prior preterm birth and a short cervix. This procedure, still in use and considered effective, involves placing a surgical stitch around the cervix to hold it closed. Unfortunately there has yet to be a head-to-head comparison of vaginal progesterone versus cerclage in effectively preventing preterm birth for women with a short cervix, so we don't know which treatment works better. Another reason why clinicians still rely on cervical cerclage may involve cost and convenience; it is a relatively straightforward procedure, and most insurance companies likely would not balk at paying for it. Progesterone treatments may be more costly. Sad, but true — cost can be a deciding factor in the therapy you receive.

And then there's bed rest. If your health care provider suspects that you have developed signs of preterm labor, the first-line therapy could be some form of this long-standing treatment, which ranges from sitting with your feet up for prescribed periods of time each day to total bed rest. When I was pregnant with my first child, I wound up on total bed rest after a routine examination at twenty-six weeks revealed that my cervix was totally effaced and two centimeters dilated; my uterus was far along in the preparations usually made for birth as a pregnancy nears forty weeks.

Does bed rest work? A recent review of published studies concluded that, decades after this practice was begun, definitive evidence still does not exist as to whether it's effective for women carrying only one baby. But in some instances, it might be quite helpful. In my case, it definitely was — my daughter was born healthy at thirty-seven weeks. In combination with bed rest, I was also told to lie in the "Trendelenburg position" — with my feet higher than my head, preferably on my side. I did this for weeks. A topsy-turvy position rolls the baby away from the weakened cervix, reducing signals that can stimulate further dilation. In addition, lying on your side increases blood flow to the uterus, which promotes growth of the baby.

Some women may be willing to opt for bed rest, especially if they have a strong support network of friends and family who can help out and an understanding employer. But it can be a tough decision for many, because of the heavy personal and physical toll that full-on bed rest involves, including its impact on your personal and professional life. For these reasons there is a tendency to avoid bed rest if at all possible.

There is a diagnostic tool that could help you and your clinician determine a course of action if you're a candidate for bed rest. The fetal fibronectin test is done between twenty-two and thirty-four weeks if preterm labor is suspected. Fibronectin is an adhesive factor, a sort of Velcro that holds the pregnancy in the uterus. This protein is normally "leaked" toward term, when the adhesive forces that bind the baby and placenta to the uterus weaken. The test, done by placing a swab in the vagina to absorb a small amount of fluid, is quick and painless, and can be repeated as necessary, if concerns about preterm birth persist.

The fluid sample is sent to a lab that rapidly determines the levels of fibronectin. If the test comes back negative, it means there is very little (though not zero) chance of delivering in the next two weeks. In other words, a negative result is good news. The test is quite accurate at making this prediction.

But there is a wrinkle. A positive result suggests there may be a cause for worry, but the test is less reliable in these cases. It has a relatively high false-positive rate, meaning that it suggests you are going to deliver soon when, in actuality, you aren't. Nevertheless, the test results can be used to determine if decreased activity or a more aggressive course of action is warranted.

It's also interesting to consider therapeutic approaches that have been tried but failed because they can offer insight concerning the mechanisms underlying preterm birth. Since infection is the most well understood cause of delivering a baby too soon, it would make sense that antibiotic therapy would be preventive. But it's not. The same goes for precautionary treatment with drugs that inhibit uterine muscle contractions. They don't work either.

The following treatments may have some benefit for the prevention of preterm labor. If you are at risk, talk to your clinician about these strategies:

* Progesterone therapy for women with a prior preterm birth or a short cervix
* Elective cervical cerclage for women with a history of second-trimester losses due to cervical incompetence or cervical dilation or shortening at sixteen to twenty-four weeks, as determined by a clinician
* Prevention, or early diagnosis, of sexually transmitted and genitourinary infections affecting the reproductive organs and urinary tract
* Cessation of smoking and use of alcohol and recreational drugs
* Prevention of multiple pregnancies as a result of fertility treatments (transfer of a single embryo)
* Daily folic acid supplementation — 0.5 mg or the amount your health care provider recommends

In addition to bed rest and the handful of other approaches already discussed, you may want to consider the following treatments and recommendations. Generally, they are thought to have either limited efficacy or no proven effect. But, as mentioned earlier, results can vary from case to case.

* Screening and treatment of asymptomatic lower genital tract infections other than STDs
* Treatment of gum disease
* Avoidance of intercourse
* Intensive education and prenatal care

Preventing Preterm Birth in the Future

When I was suddenly faced with the possibility of a preterm birth, I was shocked at how basic the preventive treatments were. At twenty-six weeks, when my "routine" pregnancy became a complicated one, I was immediately admitted to the hospital. The "prescription" at the time was wine and morphine, two substances that I would never have dreamed of taking. Now I was flying high in an effort to stop the painless contractions I had been having and had thought were normal (i.e., the harmless Braxton-Hicks contractions that many women experience). But I was willing to do anything to prevent my baby from being born three months early. Eventually I was sent home on total bed rest, and every two hours I took a drug that is now known to be as ineffective as the wine-and-morphine treatment. At thirty-seven weeks, within hours of being allowed to stand and walk, labor started in earnest, and my healthy daughter was born.

How might we change our approach to treating preterm labor that ends in birth before thirty-seven weeks of gestation? If you follow health news, you'll know that scientists are working on genetic tests for all sorts of conditions. Premature birth is no exception. Pregnancy studies are particularly challenging because they present a sort of "three body problem." In other words, the troublemaker genes could reside in the mother, the father, or the particular combination of the two that ends up in the baby. In a perfectly designed study, researchers who investigate pregnancy complications would collect and analyze DNA samples from all three individuals. But in the real world, that's unlikely to happen due to logistics and cost.

So scientists have taken a more practical approach. The inheritance patterns of preterm birth suggest that transmission most often occurs via the mother, making her DNA a good place to start the hunt for genes that might be involved. A study that took this approach analyzed data from over forty thousand women (obtained with their permission) who had personal genetic testing performed by the biotechnology company 23andme. The investigators came up with a small number of genes that might play a role in preterm birth. The hope is that one day this kind of information and a small sample of a mother's blood could be used to perform a genetic test for estimating the risk of preterm labor and birth before these processes begin. Since we know that the baby's genes and environmental factors also play important roles in this pregnancy complication, a truly effective test would also take these potential causes into consideration.

An additional reason that it's important to identify genetic risk factors for preterm birth is that the genes may lead us to the causes, which can be turned into new drug targets for developing more effective treatments. Along the way we may find important clues about what triggers the normal birth process, another mystery that needs solving.

ARE PREGNANCIES BECOMING MORE COMPLICATED?

Given all the advances in modern prenatal care, it may seem surprising that some of the most common pregnancy complications are impacting more and more women. But when you consider this fact in the context of national trends, some plausible explanations emerge.

For the past few decades, researchers have noted the rise in hypertensive disorders of pregnancy — that is, problems related to elevated blood pressure. According to the CDC, in 1993, about 5 percent of all women who had their babies in hospitals developed high blood pressure. (Hospital births are the only pregnancies that were tracked, so these figures did not include at-home births or alternative birth centers.) By 2014 that number had nearly doubled. About 10 percent of pregnant women in the study were hypertensive for one reason or another.

Why are these numbers going up? It's hard to say for sure, but some contributing factors may help to explain an overall uptick. As discussed, women are delaying pregnancy, becoming mothers when they are older. As a result, the likelihood of developing a chronic health problem that comes with age — for example, type 2 diabetes or hypertension — increases. In turn, these conditions can boost the risk of exacerbations during pregnancy or other related problems.

Another contributing factor: we are in the midst of an obesity epidemic. Recent statistics compiled by the March of Dimes show that about 30 percent of women of reproductive age are obese. (A BMI of 30 or more indicates obesity; a BMI between 25 and 29.9

indicates overweight.) This trend translates to an increased risk of complications for mother and baby.

Also, because of major advances in medical care, women with significant health problems, such as those born with heart defects or who develop autoimmune diseases such as lupus, are able to consider pregnancy, despite the fact that they are more likely to encounter some challenges and face more risks along the way. While it's good news that these women may now be able to have a baby, it also means that seeking out specialized prenatal care is a must.

Clearly, losing weight (if you are overweight or obese) and getting chronic medical conditions under control before you conceive a child are good ways to optimize your chances of having a normal pregnancy. But if we waited for all the stars to perfectly align before trying to have a baby, most of us would never become parents.

PREECLAMPSIA

Preeclampsia, which involves the new and sudden onset of high blood pressure, can turn into a serious pregnancy complication. Even for medical professionals it can be difficult to recognize in the early stages, before the full-blown signs develop. If left undiagnosed and untreated, preeclampsia can endanger the lives of both mother and baby and leave serious health complications in its wake.

Known for its sudden onset, the ancient Greeks named the syndrome "eklampsis," meaning a sudden flashing, which referred to its rapid appearance in a woman whose pregnancy, up to that point, had been regarded as normal. Seemingly out of nowhere, the end stages of preeclampsia manifest in alarming ways (such as seizures). Today we add the prefix "pre-" to signify that in most cases, with proper prenatal care, the telltale early signs can be recognized and appropriate measures taken before the situation becomes grave. Few women in the United States die of preeclampsia, but in developing countries it is one of the leading causes of maternal and infant mortality.

While most pregnancies will not be complicated by preeclampsia, the numbers are still significant. About 5 to 8 percent of all pregnancies will be affected at some level; a smaller percentage will develop into severe cases.

Hypertension: A Major Piece of the Puzzle

Preeclampsia is typically diagnosed sometime after the twentieth week of pregnancy. In the early stages of this disorder, the most common sign that something's amiss is the onset of hypertension in a woman who has never before had high blood pressure. This is why every time you visit your obstetric care provider after twenty weeks (and sometimes before), your blood pressure will be taken.

A blood pressure reading of less than about 120/80 is considered normal. The first number (systolic pressure) reflects how much pressure your blood is exerting against the walls of your arteries when your heart beats, and the second number (diastolic pressure) indicates the pressure in your arteries between heart beats. A reading of 140/90 or greater on two occasions at least four hours apart is required to diagnose gestational hypertension. A further elevation to 160/110 or above is classified as severe.

Although it's the best method that we have, there is a major problem with using blood pressure readings as a screening tool to rule out preeclampsia: the new onset of hypertension can occur for many reasons. The cause may be external, such as a stressful day. Or maybe you have "white coat hypertension," the well-documented phenomenon of feeling anxious in a medical setting at the prospect of having your blood pressure taken, which leads to an elevation only when you visit a clinic. In these cases, your blood pressure should return to normal if you simply rest for a while.

However, if in a subsequent reading it's found that your blood pressure hasn't normalized, you may have gestational hypertension. The probable cause is internal, driven by your physiology. As discussed in chapter 2, a pregnant woman's blood vessels normally relax, allowing them to carry the increased amount of blood that is required to deliver oxygen and nutrients to the baby's placenta. If pregnancy-associated widening of the arteries isn't sufficient, their internal pressure increases, and hypertension ensues.

GESTATIONAL HYPERTENSION

The new onset of high blood pressure after the twentieth week of pregnancy and in the absence of other problems results in a diagnosis of gestational hypertension. It occurs in 6 to 17 percent of healthy women who are pregnant for the first time and 2 to 4 percent of women who have been pregnant before. Many expectant mothers with gestational hypertension have healthy, uneventful pregnancies, and their blood pressure returns to normal after birth. However, if blood pressure levels reach the severe threshold (greater than 160/110), the situation becomes more serious. The baby's growth in utero could be compromised, leading to a range of problems, and a mother's health may be at risk as well.

When readings don't normalize after pregnancy, the diagnosis is changed to chronic hypertension and may need to be treated with medication if other interventions don't work. But even if readings return to normal, you now have a heightened risk of developing chronic hypertension in the future. That's why it's useful for your medical care provider to know that you developed gestational hypertension during pregnancy.

If without warning you are diagnosed with gestational hypertension, it's important to determine if this condition is occurring in isolation or if it is the first sign of preeclampsia. The urine test that is done during prenatal visits can provide additional information, though not a definitive answer. Ordinarily your kidneys filter out proteins; preeclampsia damages this filtering mechanism, allowing protein to "spill" into your urine.

* If protein is suddenly detected in your urine (without kidney disease), you will likely be diagnosed with preeclampsia.
* If protein is absent, you *may* not have preeclampsia. But this result isn't conclusive. About 10 percent of women with garden-variety preeclampsia do *not* have protein in their urine, and 20 percent of women who have the most severe forms of preeclampsia don't either. In both cases, the diagnosis is based on additional signs.

Although protein in the urine is a helpful clue, the American College of Obstetricians and Gynecologists no longer considers this marker as a requirement for the diagnosis of preeclampsia.

By now, you may have rightfully concluded that making a preeclampsia diagnosis can be quite difficult. A woman with this condition may feel perfectly well, and the problem is picked up only through screening. Moreover, different women have different combinations of its typical features. Here is a list of the main signs of impending or established preeclampsia, compiled by the Preeclampsia Foundation. Typically at least two are required to make a diagnosis.

* High blood pressure
* Protein in the urine (proteinuria)
* Swelling (edema)
* Severe headache
* Nausea or vomiting after morning sickness has subsided
* Severe pain in stomach, shoulders, or lower back
* Sudden weight gain
* Changes in vision
* Shortness of breath
* Anxiety
* Exaggerated reflexes — the muscles overreact to stimuli (to test for this, a physician uses a little rubber hammer to tap trigger points)

This is where regular prenatal care from well-trained clinicians can play a critical role. High blood pressure is often "invisible," hard to detect yourself. This makes measurements during prenatal clinic visits an imperative. Other signs, such as lower back pain or swollen ankles, can overlap with the common complaints of a normal pregnancy. Experienced practitioners have treated many pregnant women with preeclampsia, making them better able to distinguish this pregnancy complication from other bothersome but benign problems that women often experience when they are carrying a baby. Still, clinicians may sometimes miss preeclampsia because of its bad habit of hiding in plain sight. That is why it's so important for you to know what to watch for.

The Symptoms: What's Normal, What's Not

High blood pressure and protein in the urine don't sound all that serious. But preeclampsia is a process. Although it begins with hypertension and proteinuria, it can result in much more dangerous medical conditions.

Hypertension may become so severe that vital organs, such as the liver, become damaged. That is why blood tests are ordered to diagnose or rule out a very serious form of preeclampsia called HELLP syndrome. This syndrome involves **h**emolysis (breaking down of red blood cells), **e**levated **l**iver enzymes, and **l**ow **p**latelets (small specialized cell fragments that help with clotting). If preeclampsia targets the brain, a pregnant woman with no history of epilepsy may suddenly develop seizures. This is a dramatic sign that *pre*eclampsia has become eclampsia.

Some women initially develop edema, a swelling of the extremities and sometimes the face, as a symptom of preeclampsia. They also report severe headaches that do not respond to over-the-counter medications. Though it's easy to brush off these complaints as the common lot of pregnant women, you should always report them to your clinician — or call the hospital and speak to a staff member in the labor and delivery unit.

These more serious signs are less common and it is especially important to pay attention to them:

* Morning sickness is normal in the first trimester of pregnancy. Sudden nausea or vomiting occurring much later in pregnancy is not. Some experts attribute GI disturbances to liver problems or alterations in blood flow.
* Episodes of intense pain in the stomach or lower back, as well as shoulder pain, or pain while lying on your right side, may be related to liver issues.
* If you are suddenly gaining more weight than usual — for example, more than five pounds in a week — this could be a sign of edema due to damaged blood vessels that leak fluid into tissues.
* A noticeable decrease in urine production might signal abnormal kidney function.
* Blurred vision, seeing black spots or auras or flashing lights, or other changes in sight could indicate cerebral edema (brain swelling).
* A severe headache that many patients describe as "the worst in my life" could also be a sign the brain is affected.
* Getting around later in pregnancy may cause you to feel a little winded, but a noticeable shortness of breath, as well as a racing pulse and a sense of anxiety (especially if these sensations are new to you), could signal hypertension related to preeclampsia or compromised lung function.
* Some women with preeclampsia have heightened reflexes. "Hyperreflexia" may be a prelude to seizures.

Do not hesitate to immediately report these signs to your clinician.

The Preeclampsia Foundation is an excellent resource. On its website you'll find up-to-date information, short videos, and the latest research on this challenging condition. (See Resources.)

Preterm Preeclampsia

Preeclampsia can occur during the preterm period (before thirty-seven weeks of gestation) or at term (thirty-seven weeks of gestation). What causes the condition to manifest at the end of pregnancy is up for debate. Often the early-onset form is associated with shallow placentation — the placenta fails to put down roots, instead forming only

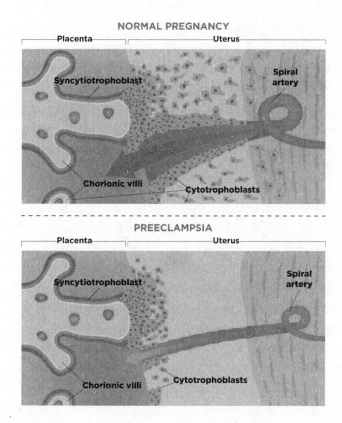

In a normal pregnancy (*upper panel*), placental cells (called cytotrophoblasts) deeply invade the uterine wall and the mother's arteries that travel through this region. As a result, these blood vessels increase in diameter, which enables delivery of an adequate amount of blood to the placenta. In early-stage preeclampsia (*lower panel*), invasion of the uterine wall and blood vessels is shallow. Many of the mother's arteries in that region remain the same as they were before the pregnancy. They are too small to deliver enough of her blood to the placenta and the baby.

superficial attachments to the uterine lining and its blood vessels. Normally, the organ deeply invades the uterus, tapping blood from its enlarged arteries (see chapter 2). But in this subtype of preeclampsia, the shallow invasion means the placenta and developing baby don't get an adequate supply of nutrient- and oxygen-rich maternal blood.

The big mystery is how faulty placentation leads to the maternal signs of preeclampsia. There is no doubt that the placenta is the source of the problem. On rare occasions a genetic defect causes cytotrophoblasts to form a benign uterine mass in the absence of a baby. If undiagnosed, a significant number of women with this condition will develop preeclampsia, proof that you need only a malfunctioning placenta — and not the child it supports — to develop this pregnancy complication.

One theory is that haywire placental cells release toxic substances into the mother's blood. This is why preeclampsia was once called toxemia (blood poisoning). Work from our lab and from other investigators has implicated molecules (called sFLT1 and PLGF) with potent effects on blood vessels. In preeclampsia, their levels slip out of the healthy range, and vascular damage ensues. But other harmful substances, yet to be identified, must also be playing a role. No matter their identity, it's clear that their target is the mother's blood vessels, where they cause moderate to severe damage; this is reflected in the wide range of forms that the clinical presentation of preeclampsia can take. This explains the high blood pressure that can manifest as a severe headache. The damaged blood vessels may also "leak," causing fluid accumulation in the tissues (edema). It also explains why protein suddenly spills into the urine from the kidneys' damaged vasculature. In the worst-case scenario, blood vessels can rupture, causing damage to the organs they supply. The brain and the liver are particularly vulnerable to this type of injury.

Over the past twenty years, the rates of preeclampsia in all its forms, occurring during the preterm period or at term, have risen by 25 percent. No one fully understands why, but one explanation may lie in the factors that heighten a woman's risk of developing this pregnancy complication. Obesity, preexisting hypertension, and diabetes (types 1 and 2) are risk factors, and the incidence of these conditions has been rising in recent years. The fact that women are waiting until they are older to start a family may be relevant, since advanced age (greater than thirty-five years) can predispose a woman to this condition. Medical professionals and researchers largely agree that the following conditions are risk factors.

* Chronic hypertension
* Antiphospholipid antibody syndrome (an autoimmune disease)
* Systemic lupus erythematosus
* Pre-gestational diabetes
* Chronic renal disease
* Carrying multiple babies
* Pre-pregnancy BMI greater than 30
* Previous stillbirth
* First pregnancy (nulliparity)
* Maternal age greater than 40
* Long interval between pregnancies (greater than five years)
* Reduced school education
* Previous preeclampsia
* Assisted reproduction
* Previous intrauterine growth restriction
* Previous placental separation before birth (placental abruption)

Some of these risk factors provide hints about the poorly understood mechanisms that drive the development of preeclampsia. A family history of this condition raises your likelihood of having it, as does being diagnosed with preeclampsia in a previous pregnancy. Both observations suggest that, in some cases, there is an as-yet-unidentified genetic component.

Having an autoimmune disease, such as lupus or antiphospholipid syndrome, is also a risk factor, possible evidence that an overactive immune system plays a role. Carrying multiple babies predisposes a woman to preeclampsia, suggesting that there is a limit to the amount of placental tissue that a woman's body can accommodate. Any condition that impairs kidney or cardiac function also heightens risk, probably due to the body's inability to respond to the added demands of supporting itself and a baby during pregnancy. Clotting disorders can also predispose a woman to preeclampsia, perhaps because they compromise blood flow to the placenta. It's harder to suggest reasonable rationales for other risk factors: in vitro fertilization, non-white race, lower educational level, and being pregnant for the first time.

Although clinical signs of preeclampsia are most frequently detected in the mother, the baby may also be affected. The primary problem is caused by the reduction in blood flow to the placenta. With fewer available nutrients, growth slows down. Intrauterine growth restriction (referred to by the acronym IUGR) may be the first sign of preeclampsia. The diagnosis is made by ultrasound imaging. With today's technology it is possible to very precisely determine a baby's weight before birth.

Another mysterious thing about preeclampsia is that sometimes it does *not* occur. Prenatal visits to your caregiver may reveal that the baby's growth has slowed or even stopped, and it may be due to shallow placental invasion that for all the world looks just like what happens in preeclampsia. But some women seem resistant to the effects of a malfunctioning placenta. The maternal signs never materialize. The problem manifests with the baby.

Birth and removal of the placenta do not end the risks of preeclampsia. A patient's symptoms may not abate until several days after delivery. Or a woman who experienced no signs of this complication during pregnancy may find that the symptoms appear for the first time up to six weeks after delivery. In this latter case, it may be that lingering damage to the blood vessels, which occurred during pregnancy, now rebounds, producing hypertension and other problems, such as visual disturbances or seizures. These symptoms are evidence of compromised blood flow to the brain, the organ that is apparently most vulnerable to the effects of postpartum preeclampsia, as this condition is called.

And the underlying problems that may have brought on preeclampsia do not disappear after the baby is born. Many studies have shown that women who had preeclampsia have a greater risk of developing various forms of cardiovascular disease as they age. Therefore, it's important to let your regular health-care provider know if you had preeclampsia. You will then be more closely monitored for conditions such as hypertension. Despite the link to cardiovascular disease later in life, there is also some good news. Many studies have reported that among women who had preeclampsia the risk of breast cancer decreases — another enigmatic observation that researchers think might eventually provide insights into both conditions.

Prediction, Prevention, and Treatment of Preeclampsia

Given the potential dangers of preeclampsia and the difficulty of diagnosing this complication, it would be ideal if a simple test, performed early in a pregnancy, could identify those women with the highest risk of developing this condition. Unfortunately, none of the current imaging or blood tests are very good predictors. That's why the American College of Obstetricians and Gynecologists (ACOG) recommends that risk levels be assessed in a thorough medical exam in the first trimester, to screen for the factors that may predispose a pregnant woman to preeclampsia. If you are at higher risk, you and your obstetrical care provider can be especially vigilant about possible signs, particularly the onset of hypertension.

Preventive strategies mostly involve taking over-the-counter medications thought to be safe for pregnant women; they have been widely used with no apparent serious side effects. The strongest specific recommendation comes from ACOG, which advises that high-risk women should take a low dose of aspirin (81 mg) every day after the twelfth week of pregnancy. The drug, an anti-coagulant, may prevent clots from forming, in order to keep blood flowing from the uterus to the placenta. Trials of calcium supplementation showed some benefit but only among populations with low levels of calcium intake through their regular diets or supplements — not usually a factor in the United States or other developed countries. Administration of vitamin C or E, both of which are antioxidants, has been attempted because oxidative stress is thought to be a component of preeclampsia pathology. But this potential means of prevention failed.

Treatment of preeclampsia depends on how far along you are in pregnancy when a diagnosis is made. Near term, thirty-seven weeks and beyond, the common approach is delivery. The only known definitive cure for this pregnancy complication is removal of the malfunctioning placenta, which means that the baby has to come out too. After birth, the signs of preeclampsia usually disappear quickly. But as described above, there are exceptions.

If the diagnosis is made before thirty-four weeks of gestation, the approach is usually what practitioners call "expectant management" — careful watching and waiting. In the United States, the basic elements usually include reduced activity and observation, as well as very close monitoring of the signs (including blood pressure, protein in the urine) either at home or in a hospital. In the UK, hospitalization is mandated, and rates of maternal mortality from preeclampsia there are the lowest in the world. In cases where blood pressure is dangerously high and stays that way (greater than 160/110), anti-hypertensive medications are often given. If preeclampsia progresses to the point where it is unsafe for the pregnancy to continue, the only option is removal of the placenta, which necessitates delivery, even if the baby is premature. In this case steroids are often given in anticipation of birth. They cross the placenta and protect the baby from lung, brain, and intestinal problems. You may also receive magnesium sulfate, which protects the brain, warding off seizures.

The gray zone is when preeclampsia is diagnosed between thirty-four and thirty-seven weeks of gestation. The natural inclination is to delay delivery, giving the baby a chance to mature further. However, delivery may be necessary if the mother's symptoms worsen and the condition progresses. Furthermore, the stress of preeclampsia can halt a baby's growth, removing the rationale for waiting.

Managing Preeclampsia in the Future

I predict the eventual development of a blood test, based on levels of molecules with causative roles, for identifying women who are likely to develop preeclampsia. The need is great, particularly given the rising numbers of cases and the difficulty in diagnosing this complication. Even when women and their caregivers are familiar with the early warning signs of preeclampsia, making a diagnosis can be challenging, because so many of its symptoms are easily mistaken for common complaints in pregnancy or chronic medical problems. Having a simple, specific diagnostic test that works before preeclampsia can take hold and worsen would make pregnancy a much safer process. A definitive negative test would also provide enormous peace of mind to pregnant women who now have to undergo repeat testing to rule out this condition — a situation that also drives up the cost of prenatal care.

Another reason that I think a blood-based diagnostic test for preeclampsia will eventually become a reality is the many well-established precedents for other medical conditions. If liver disease is suspected, then a blood sample is used to determine if this organ is functioning abnormally. This works because the liver — a veritable factory — is bathed in blood, and some of its products end up in the bloodstream. When this organ malfunctions, its production line goes berserk. It makes too much of this and too little of that. The same relationship between organ and blood holds true for the placenta. We already know that a normally functioning placenta releases a vast array of substances into a mother's circulation. When a placenta malfunctions, as in preeclampsia (and other pregnancy complications), these substances must change. We just need to do the research to pinpoint the precise differences.

GESTATIONAL DIABETES

Diabetes occurs when levels of glucose (sugar) in the blood are higher than they should be. Glucose provides energy to every cell in the body, a process regulated by the hormone insulin, which is produced by the pancreas. Insulin acts like a crossing guard, directing the trafficking of glucose from the blood into cells, where it is turned into energy or stored.

Type 1 diabetes is an autoimmune disease. For poorly understood reasons, a person's immune system becomes confused and kills the insulin-producing cells of the pancreas as if they were disease-causing organisms, such as bacteria and viruses. In the absence of insulin, glucose can't enter the cells where it's needed for energy, and blood glucose levels rise. Diabetes causes a broad spectrum of health problems because high levels of blood glucose act as a poison.

Type 2 diabetes results from insulin insufficiency. The pancreas may have an inherited limitation in its ability to produce this vital hormone, or it can't match the increased need for insulin brought on by weight gain, inactivity, and aging. The result is the same as in type 1 diabetes — rising blood glucose levels and sugar-starved cells.

Gestational diabetes is a specific form of this disease, which appears during the second or third trimester of pregnancy and usually disappears after delivery. (Elevated blood glucose levels *before* the second trimester probably indicate a preexisting condition; that is, the mother was diabetic before she became pregnant.) During pregnancy, chronically high blood glucose levels can damage your cells and those of your baby.

Many women with gestational diabetes have no symptoms or mild ones easily attributed to being pregnant, such as increased thirst and frequency of urination. And today few pregnant women have no risk factors for gestational diabetes. This is why testing, usually between the twenty-fourth and twenty-eighth week of pregnancy, is a standard part of obstetric care in the United States and other developed countries. If your clinician is concerned that you have diabetes, you may be tested earlier.

Here are the most common risk factors and symptoms associated with gestational diabetes, compiled by ACOG, which offers detailed information for women with diabetes. (See Resources.)

Risk Factors

* Age forty-five or older
* Being overweight or obese
* Family history of diabetes
* Physical inactivity
* Particular racial or ethnic background: Native American, Asian, Hispanic, Black, Pacific Islander
* Previous abnormal glucose-screening results
* High blood pressure
* High cholesterol
* History of gestational diabetes or a baby weighing more than nine pounds at birth
* Polycystic ovary syndrome (PCOS)
* History of cardiovascular disease

Symptoms of Type 1 Diabetes

* Increased thirst or urination
* Constant hunger
* Weight loss without trying
* Blurred vision
* Extreme fatigue

Symptoms of Type 2 Diabetes

* Any of the symptoms of type 1 diabetes
* Sores that are slow to heal

* Dry, itchy skin
* Loss of feeling, or tingling, in the feet
* Infections that keep coming back, such as yeast infections

Two tests are used to check for gestational diabetes; there is not enough evidence to prove that one is better than the other. Both involve drinking a specially formulated glucose-containing beverage, after which one or more blood samples are taken and glucose levels measured.

The first, called a glucose challenge test, is for screening purposes, and is read as an initial evaluation; it determines the likelihood of your having diabetes. It's quick and easy compared to the second test because no preparation is needed. You drink the cocktail, and an hour later your blood is drawn and the glucose level measured. If it's less than 140 mg/dL (140 milligrams of glucose per deciliter), the likelihood of diabetes is low. If it's 200 mg/dL or more, you will probably receive a diagnosis of gestational diabetes.

If the level is between 140 and 200 mg/dL, you will need a follow-up diagnostic procedure called an oral glucose tolerance test. Some health care providers skip the screening step and go straight to this diagnostic test. It is more involved because it requires fasting for eight hours (nothing to eat or drink except water) before you go to the clinic for an initial blood draw. There, you drink the glucose-containing cocktail, and your blood is drawn every hour for two to three hours. High glucose levels at any two points in time are diagnostic for gestational diabetes.

The prevalence of gestational diabetes among expectant mothers in the United States is estimated at 6 percent. Around the globe, numbers fluctuate greatly among the different populations, perhaps in some cases reflecting the variety of testing strategies; the range is 1 to 25 percent. Further complicating these matters, experts have failed to reach a consensus on the blood glucose levels that indicate a diagnosis of gestational diabetes. It's good to be aware of this discrepancy. Based on the same test results, one clinic might give a woman a diagnosis of gestational diabetes, while another clinic would not. Adopting more rigorous criteria across the board would no doubt increase the percentages of women diagnosed with this condition.

To understand the reason for disagreement, think about the idea of a "safe" level of alcohol consumption during pregnancy, discussed earlier. Is one drink okay? More? None at all? With alcohol, eliminating consumption altogether during pregnancy unquestionably improves outcomes, but you will still hear different opinions about what's okay. Similarly, with blood glucose, the "safe" level above a normal range is hard to determine absolutely. But here is a helpful fact: the higher a pregnant woman's blood glucose levels, the higher the risk of a pregnancy complication. So, even for women with mildly elevated blood glucose, outcomes will improve if the level is lowered. Therefore some clinicians choose to err on the side of safety and make a diagnosis of gestational diabetes even when blood glucose levels are lower than the standard cutoffs.

One thing researchers do agree on is that gestational diabetes can elevate the risk of certain pregnancy complications, including these:

Problems for the Mother

* Preeclampsia
* Too much amniotic fluid (hydramnios)
* Complicated vaginal delivery, possibly leading to cesarean delivery, due to the baby's higher-than-normal birth weight
* Birth trauma to the mother due to the increased size of the baby at delivery

Problems for the Baby

* A small but definite increase in mortality just before or after birth

* Respiratory problems after birth
* Metabolic complications: low glucose levels, low calcium levels, jaundice
* Abnormal increases in blood volume and number of red blood cells
* Enlargement of organs such as the liver and the heart
* Birth trauma to the baby due to increased size at delivery

The Role of the Placenta in Gestational Diabetes

During pregnancy, a complex balancing act maintains healthy glucose levels in a mother's blood and that of her baby. The placenta is an important control point. As discussed in chapter 2, this involves higher amounts of glucose in the blood of the mother than in the baby's. Imagine a drop of food coloring diffusing throughout a glass of water. The color fades as it spreads. This is what happens with glucose as it moves from a mother's blood, where the concentration is higher, across the placenta to reach the baby's circulatory system, where the levels are lower.

Maintaining this balance entails coordinating the pancreatic functions of baby and mother. During early pregnancy, the baby's pancreas, which is still developing, secretes lots of insulin, keeping glucose levels in fetal blood low relative to that of the mother. Then comes the twist. In response to the growing baby's energy needs, the placenta carries more and more glucose to the baby, lowering levels in maternal blood and potentially upending this balance. Ever resourceful, the placenta tries to solve the problem it creates by secreting hormones that make the mother mildly diabetic. Her blood glucose concentration rises so that she can maintain the appropriate level in her baby's circulation.

Then something truly amazing happens. Pregnancy causes the cells of the mother's pancreas to increase in number and secrete more insulin. The added dose of insulin "cures" the diabetes, maintains the glucose levels in maternal blood, and ensures that the baby receives just the right amount needed for normal growth.

In gestational diabetes, the mother's pancreas does not adequately respond to the signals that tell it to rev up. As a result, her blood glucose levels remain abnormally high, which increases placental transfer of this sugar to her developing child. It's as if the baby was being given a sugary drink to consume. This is why babies of mothers who have gestational diabetes and poorly controlled glucose levels weigh more at birth than they should, complicating delivery and raising the rates of surgical interventions. These newborns can also run into metabolic problems as they adjust to the world beyond their mother, where there is a sudden drop in glucose levels.

Treatments

If you receive a diagnosis of gestational diabetes, you will be asked to monitor your blood glucose levels, using a meter that analyzes a small drop of blood from your finger, and to record the values. The NIH recommends that women aim for target blood glucose levels calibrated to different times of day. (See Resources.) An hour after eating, the goal should be 140 mg/dL or less, dropping to 120 mg/dL at the two-hour mark. At other times — before eating, bedtime, and overnight — the target is 95 mg/dL or less.

Your clinician will work with you to formulate a plan for controlling your blood glucose levels. We live in the age of personalized medicine, which acknowledges that everyone is different and a one-sized approach does not fit all women with gestational diabetes. The particulars will vary, but most plans will have three basic elements: diet, exercise, and, if necessary, insulin treatment.

Dietary changes can be highly effective. By altering what you eat, you can lower your blood glucose levels and limit the amount of sugar the placenta transports to the baby. It may be that a few changes, such as eliminating fruit juice or sugary breakfast cereal, could make a big difference. Consultation with an obstetric dietician can be very helpful.

That said, even under normal circumstances, it's hard for most of us to fundamentally change the way we eat. To receive a diagnosis of gestational diabetes and discover that your normal pregnancy has suddenly become a high-risk one is extremely stressful. Feeling anxious is a normal reaction, and it may be difficult to drum up the motivation to overhaul your diet. But it's worth it! Most women with gestational diabetes can control their blood glucose levels with straightforward dietary measures. If you try, you have a good chance of succeeding.

The typical strategy is to significantly reduce carbohydrate intake, but it can be hard to adhere to this regimen. Many women find a very-low-carbohydrate diet too restrictive. Some, in response to this challenge, substitute fats for carbohydrates, which doesn't help. Elevated levels of fat in the blood promote insulin resistance, which raises blood glucose levels and causes the placenta to transfer excess amounts of glucose *and lipids* to the baby. This compounds the problem, and the baby will continue to grow too fast.

Some experts are proposing an alternative approach — not only reducing but also changing the types of carbohydrates that you eat. Foods containing sugar (desserts and candy) are decreased, and higher-quality complex carbohydrates — starches and fiber-rich foods such as whole grains, beans, and vegetables — are increased. If you normally have problems (such as gas and bloating) when you digest high-fiber fare, you'll be more successful if you gradually increase your intake of these foods a little each day.

There is a growing awareness that in order to succeed, a dietary plan must be sensitive to ethnic and regional eating habits. Most of us have a very deeply rooted fondness for the foods we grew up with, and they can be very hard to resist. Any diet plan should acknowledge this fact while making useful adaptations. If rice is a staple in your diet, substituting brown rice for white rice might work for you.

Or you could add beans to the rice, which results in better control of glucose levels.

Physical activity is another important element of an effective treatment plan. However, it is crucial to first talk to your obstetric care provider to find out if you have one or more of the risk factors that make exercise during pregnancy inadvisable. (For information on exercise, see chapter 6.) If you have gestational diabetes, physical activity becomes even more important. The NIH advises aiming for thirty minutes of exercise five times a week. If this is already part of your regimen, you can discuss with your care provider the option of ramping up the intensity, something you should not do without having this important conversation.

If diet and exercise are not effective in lowering your blood glucose levels, the next step may be insulin injections. There are other, newer drugs to treat diabetes, but to date there is not definitive data about using them during pregnancy, so insulin remains the standard treatment. Insulin is proven to be effective, and it's safe for the baby because it does not cross the placenta.

Long-Term Effects on Mother and Baby

In countless ways, pregnancy is a stress test. Gestational diabetes could be a sign that your body was struggling to keep blood glucose levels at a normal range even *before* pregnancy. The added work of supporting a baby's growth can stretch your glucose control systems beyond the breaking point. This helps explain why women with gestational diabetes have at least a sevenfold increase in the likelihood that they will develop type 2 diabetes during the next decade.

The good news is that this heightened possibility comes with an expiration date. If after ten years you don't develop diabetes, you can consider your risk to be the same as the average person's. Losing weight and exercising can help you reach this milestone.

The fact that many women with gestational diabetes never again have problems keeping their blood glucose levels in a normal range suggests that they could have a different problem that does *not* predispose them to developing type 2 diabetes. This group might not adequately respond to the signals that change how a mother's blood glucose levels are regulated during pregnancy. After birth, this regulatory system is no longer needed, and the problem disappears forever, or at least until the next pregnancy, when it might recur.

Gestational diabetes also has consequences for the baby (to be discussed in chapter 6), whose risk of eventually developing type 2 diabetes increases. Fortunately, we know that management of the mother's gestational diabetes reduces the risks of adverse outcomes for her baby, and a healthy diet and exercise can lower the lifetime risk for the child.

Managing Gestational Diabetes in the Future

Scientists, including my group and our colleagues who are members of the Diabetes Center of the University of California San Francisco, are very interested in the placental signals that tell the mother's pancreas to make more insulin-producing cells during pregnancy. Identifying these messengers will enable the development of synthetic versions, which will mimic their effects. This powerful new class of drugs could be used to treat various forms of diabetes.

Some cases of gestational diabetes — in particular, those in which the woman does not develop chronic diabetes after pregnancy — likely involve faulty communication between the placenta and the mother's pancreas. If we could intervene and somehow correct the signals, an effective cure might be possible, a way to treat the cause rather than the symptom of high blood glucose levels.

In type 1 diabetes, the pancreas loses its insulin-producing cells. Just the opposite happens during pregnancy. Placental signals expand this population. It's likely that this mechanism has broader utility, beyond pregnancy, and for men as well as women. If true, these signals might be able to reverse the loss of insulin-producing cells that leads to type 1 diabetes, the form that most commonly occurs in children and young adults. Given the large increases in the incidence of diabetes worldwide, new therapies are urgently needed. Learning whether we can co-opt this mechanism to stave off other forms of diabetes is another exciting direction in research.

RIGHTING PREGNANCY WRONGS

As I know from my own experience, facing a serious pregnancy complication can be one of life's hardest challenges. But even then, you *can* take charge by having well-researched information close at hand that you can use to make informed decisions.

In medicine, understanding how something goes *right* is often the key to understanding why things sometimes go *wrong*. I have been studying the causes of pregnancy complications for decades, but that work grew out of the research my group was doing on normal placental development and function. By the late 1980s, my research team and I had figured out some of the molecular drivers that make for a healthy placenta, a prerequisite for a successful pregnancy. As it turned out, we were able to use our discoveries to zero in on failures in placental development and how they contribute to preterm birth, preeclampsia, and gestational diabetes. Eventually, we hope to use our discoveries to correct these missteps earlier in pregnancy, long before the signs appear, so that we can set mother and baby back on the right path.

As discussed in the next chapter, new research suggests there are many reasons, beyond the desire for a healthy baby, to prevent pregnancy complications. Surprisingly, a poor uterine environment may have a negative impact on a child's future health that will manifest years later, in the form of chronic diseases that strike during adulthood. But fortunately, researchers are also providing us with strategies for preventing or reversing these unanticipated consequences.

6

How Life in the Womb
Determines Adult Health

Before your baby is born you'll make many decisions, all with the goal of bringing a healthy child into the world. But the impact of these choices goes beyond birth. When you take charge of your pregnancy, it turns out you're also taking charge of your child's health — and not just at birth, in infancy, and in early childhood. Mounting evidence points to the fact that a baby's life in the womb can be a predictor of his or her health *as an adult* — a fascinating finding that provides a strong rationale for optimizing your health during pregnancy. At the same time, you'll be providing the best possible environment for your baby's development, which has the added benefit of lowering your child's future risk for the diseases that become increasingly common as we age.

Welcome to the world of "fetal programming of adult health," as it's known in the research community. The phrase may sound like it belongs in a work of science fiction rather than a book on pregnancy. But fetal programming is real, a revolutionary concept that is rapidly changing how we gauge the importance of the nine months a baby spends in the womb.

Very recently scientists have made the surprising discovery that our journeys into this world, from conception through birth, can factor into our health as adults. If the trip goes smoothly and a baby arrives on time (and not too early), a normal pregnancy can lay the foundation for a lifetime of good health. If, on the other hand, a baby's prenatal period is a rough ride, compromising growth before birth, then the opposite might be true: that individual may have a higher risk of developing certain common medical problems, such as cardiovascular disease and type 2 diabetes, and these conditions won't show up until adulthood.

Fetal programming of adult health may sound eerily like predestination — that fate has the upper hand, and there is nothing you can do to change a baby's health once his or her life course is established in the womb. But we now know with certainty that women can take many steps before, during, and after pregnancy that will optimize the chances of having a baby who grows normally before birth, through the earliest years of life, and beyond. You've read about the importance of preparing for pregnancy, reducing your exposures to environmental toxins, and lowering your risk factors, where possible, for certain complications. In this chapter, you'll complement that knowledge with even more practical take-charge steps.

But before we get to that information, let's take a moment to understand *why* a baby's life in the womb and adult health are inextricably linked. It took decades to trace the origins of this relationship, but ultimately scientists were able to make this vital connection. They didn't do it with fancy diagnostic tests or high-tech tools. Instead, they reached back in time and studied the life course of people born more than a hundred years ago. These adults were once newborns. And their medical histories provide us with invaluable lessons about optimizing pregnancy and the health of babies born today, including your own.

WHAT THE BABIES TAUGHT US

In the mid-1980s, the UK researcher Dr. David Barker, who trained as a physician and epidemiologist, published his research on the link between fetal health and adult disease. His discoveries were startling.

Barker was able to connect malnutrition in infancy and childhood to heart disease in adulthood by looking at mortality trends across England and Wales from 1921 to 1925. Remarkably, areas that once had high rates of infant mortality now reported the most deaths due to heart disease. Barker and his research partner, the statistician Clive Osmond, inferred that these deaths were somehow linked to economic hardships in the affected areas, where nutritious food would have been scarce when these adults were babies and children — and when they were in the womb. After reviewing the data, the two concluded that adults born in areas with limited access to food who later in life began to eat the standard high-calorie Western diet (with more fat and sugar than they'd been used to) had an elevated risk of heart disease. The Barker hypothesis, concerning the developmental origins of adult health and disease, was coming into focus.

This was not an entirely new idea. In the first decade of the twentieth century an American pediatrician began to suspect that what an infant ate influenced metabolism later in life. Dr. Henry Dwight Chapin spent decades treating neglected, malnourished children from the slums of New York City who had been placed in institutions where conditions were not much better. He was among the first to conceive of a system of "adopting out" young children to foster homes, where they could live a more normal life that included better nutrition. Dr. Chapin saw thousands of orphans blossom into healthy children once they were placed in stable homes. He closely followed the progress of the children, but over time he realized that deprivations early in life could have a long-term impact, somehow materializing many years later as poor adult health.

Now Barker and Osmond had hard numbers and data that supported Chapin's observations, but they needed even more evidence to push the theory forward. Skeptics questioned whether their findings were purely coincidental, a statistical fluke. Others would not believe fetal programming existed until the researchers could provide a biological mechanism that explained the link between poor nutrition in early life and heart disease in adulthood. Others wanted to know if genetics were the culprit. Barker was trying to advance his ideas in the mid-1980s, when the rush was on to find genetic causes for common diseases.

Eager to answer his critics in the best way a researcher can — with more data — Barker reasoned that whatever caused the high rates of infant mortality must also have had a negative impact on birth weight and subsequent growth during early childhood. But in order to test this theory he would need three key pieces of information for every individual in his study: birth weight, weights in early infancy, and documentation of the person's adult health. Gathering these data in his native England was hard work that took many years, but thanks to the national health-care system and its records, it was not impossible.

Barker eventually tracked down meticulously compiled ledgers that documented the birth weights and growth of thousands of babies during their first year of life — 5,654 males born between the years 1911 and 1930. Ultimately, he and Osmond were able to match them to the health records of the men these babies became. What they found was an exciting confirmation and extension of their original data. Those with the lowest weights at birth and at one year had the highest rates of death due to heart disease. The researchers concluded that "environmental influences that impair growth and development in early life" — including malnutrition — were probable risk factors for deadly heart disease.

The implications of the new data were profound. Poor prenatal and postnatal growth set the stage for heart problems later in life. But this broadening of the original theory once again generated criticism and controversy. Had the researchers taken into account changes in other environmental factors that occurred between the birth of the under-weight babies and the onset of heart disease decades later? What about shifts in diet, adoption of a more sedentary lifestyle, or the increase in popularity of cigarette smoking? And, given that Barker and Osmond's data focused only on British males, what about health patterns in other populations? Would these ideas hold up for men and women around the world?

The answer, as it turns out, is yes. Barker and Osmond were no longer alone in researching questions about the link between fetal and adult health. Around the world, many groups jumped into the fray and began to replicate and extend their studies. For instance, researchers in India, focusing on children at eight years of age who had had a low birth weight, found they had higher than expected blood pressures, serum lipid profiles portending cardiovascular disease, and evidence of developing diabetes. This study, whose outcome the Barker hypothesis predicted, was important for another reason. It suggested that investigators did not have to wait through the course of a person's lifetime to find evidence of the adverse health effects of low birth weight or poor growth during infancy. The evidence was apparent during childhood.

Eventually the developmental origins of health and disease would be confirmed by numerous studies in racially and ethnically diverse populations worldwide. Whatever our differences, this vulnerability seems to be something we humans share — what happens while we're developing in utero and during infancy and childhood can affect our health decades later.

Thrifty Babies: From Famine to Feast to Fat

The explanation of the link between malnutrition in early life and heart disease later on may lie in a subsequent theory pioneered by Barker and another colleague, Charles Nicholas Hales. Known as the "thrifty phenotype hypothesis," it goes like this: when experiencing poor nutrition during pregnancy and early childhood, the baby in the womb and newborn infant learn to be "thrifty" with the limited nutrients they are given, essentially altering their metabolism to conserve precious resources and expend less energy.

This was likely a survival strategy that evolved because food has been scarce for most of human history. Exposure to poor nutrition during this critical window somehow molds an individual's basic physiology. If their nutritional status does not change dramatically, the status quo is maintained and health is relatively normal (excepting any serious nutrient deficiencies, which could lead to other negative health consequences). But if the same individual suddenly begins eating a rich diet, their inability to respond to this metabolic challenge can elevate the risk of cardiovascular disease and type 2 diabetes, caused by a kind of metabolic overload.

Here is another way to think of the adult metabolism of a formerly undernourished infant. Imagine the baby as a small, highly efficient power plant that has learned to run very effectively on minimal resources. Suddenly, the plant receives a massive influx of power. There is more fuel coming in than it can turn into energy! The only thing to do is process the raw material into a form that can be put into long-term storage in case scarcity returns. By analogy, a "thrifty" baby's sudden exposure to a rich diet can cause metabolic chaos. Food consumed over and above what is needed for energy production gets stored as fat, leading to obesity, cardiovascular disease, and diabetes.

Despite these breakthroughs in understanding, many questions remain. Are cardiovascular disease and diabetes the only medical problems linked to fetal programming? Does poor growth during pregnancy and early childhood increase the risk of other common diseases, including cancer? Researchers all over the world are trying to answer these questions. In most cases, the jury is still out, but as research continues, investigators continue to make surprising findings.

UNDERSTANDING OVERNUTRITION

Overnutrition is not the same as overeating, though in both cases a person may be consuming more calories than she requires. Specifically, overnutrition means that you are taking in more of a particular nutrient than your body needs. If you eat a balanced diet and you're healthy, but you consume a little more of a vitamin or mineral than you need, your body can usually handle it by eliminating the surplus. For example, you may be vacationing in a place where citrus fruit is in season, and you can't stop eating delicious, just-picked oranges and drinking freshly squeezed OJ. You are probably taking in more vitamin C than you can use. You may get an upset stomach, but you'll be okay because your body can deal with the extra amount. Worse would be if you ingested excessive iron — a mineral the body can't easily eliminate, and which can be toxic at high levels.

In fetal programming, the real culprits aren't vitamins and minerals. Instead, overnutrition refers to too many saturated fats and simple carbohydrates in the diet, which can alter metabolism — including that of a developing baby in the womb — and ultimately heighten the risk of developing obesity, heart disease, and type 2 diabetes in adulthood (and according to some studies, childhood). Bottom line? The "nutrition" in overnutrition is rarely nutritious and is never a good idea, whether or not you are pregnant.

How Maternal Overnutrition Can Harm a Baby's Adult Health

During the first half of the twentieth century, many countries, including those that Barker and Osmond studied, were plagued by a lack of food. Two world wars within one generation contributed to famine, disease, and other hardships that took a heavy toll on human health. But as the world emerged from this conflict and an economic depression, commercial food producers figured out ways to increase output and reach more consumers. As a result, the landscape of available food began to change.

By the mid-twentieth century, people in these same regions were rapidly gaining easy access to many food products that were affordable and convenient but high in fats and simple carbohydrates such as sugar. Even in rural areas, cheap processed food abounded. People had more to eat, but it wasn't necessarily nutritionally dense, healthy food with a favorable nutrient-to-calorie ratio.

The trend toward more industrialized food supplies coincided with other factors that would eventually have an impact on health. By midcentury, the nature of wage-earning work was changing for many — from physical labor to sedentary office jobs — and cars were everywhere. Much of the population became less physically active. More food, less exercise . . . the stage was set for metabolic calamity.

Americans, in particular, began to eat more animal protein, fewer plants, and more fat — especially artery-clogging trans fat, which is found in fried foods and food products made with hydrogenated and partially hydrogenated oils. Researchers now know that trans fat is detrimental to cardiovascular health and is probably a major contributor to today's skyrocketing rates of obesity, even among children. Obesity is now a global epidemic, occurring in both industrial and developing countries.

In a relatively short time, concern about undernutrition of reproductive-age women was replaced with overnutrition, which became a major problem. Researchers began to investigate the effects of obesity on pregnancy, and what they found was right in line with the Barker hypothesis.

When scientists studied the babies of overweight and obese mothers, compared to those whose mothers had normal weights, they discovered evidence of fetal programming. Often the effects were hard to untangle from those of diabetes, which can be prevalent among this group of mothers. A Finnish study focused on the characteristics of those who gave birth in 1986 and the health of their children when they reached adolescence. If a woman was overweight (having a BMI greater than 25) before and during pregnancy, her child was more likely to be overweight with abdominal obesity (excessive fat around the stomach) at age sixteen. The risks were even higher when the overweight mother had diabetes. Another study found a similar association between elevated maternal weight and a child's risk factors for developing cardiovascular disease, diabetes, or both as a young adult.

Researchers have also studied fetal programming in response to overnutrition caused by maternal diabetes. Diabetic mothers, even those who are not overweight or obese, have elevated levels of nutrients in their blood, which are delivered to the placenta and baby. To address only the effects of diabetes, one group of researchers studied a sibling born to a normal-weight mother before she was diagnosed with type 2 diabetes and a sibling born after the fact. The risk of developing diabetes (and obesity) was significantly higher for the sibling who was born after the diagnosis and prenatally exposed to a diabetic environment.

With the results of these studies in mind, it's easy to see why avoiding overnutrition and lowering your risk factors for diabetes doesn't just benefit your own health — it can also protect your child from cardiovascular disease, diabetes, and weight gain that can lead to obesity as he or she grows into adulthood.

THE PLACENTA'S ROLE IN FETAL PROGRAMMING

What sort of biological maze starts with a mother's under- or over-nutrition during pregnancy and ends years later, when the child, now grown into an adult, develops a medical condition? Clearly some of the paths in this complex labyrinth lead to dead ends, because not every baby who grows poorly before birth, or every newborn who fails to thrive, is destined to develop a chronic disease in adulthood. It's the *routes* that lead to cardiovascular disease, diabetes, and perhaps other chronic health problems that scientists are trying to figure out. And placental development and function may hold some answers.

Researchers are exploring a possible role for the placenta and the maternal nutrients it delivers to the baby. Dr. Barker was avidly pursuing this line of inquiry when he suddenly died in 2013. In an article written near the end of his life, he suggested that "variations in the process of placental development lead to variations in the supply of nutrients to the fetus," which would, in turn, "program" the developing baby in ways that could lead to chronic disease later in life.

This theory makes a great deal of sense. As discussed in chapter 2, one of the placenta's most important jobs is to provide adequate nutrition to the growing baby. Its failure to do so could kick off the chain reaction that leads the baby to become "thrifty" and sets the course for fetal programming of adult diseases.

How might this happen? In general, placental size correlates with that of the baby. Big babies have big placentas and small babies have small ones. But this does not always hold true. There is great variability in placental size among babies, even those of the same weight. What happens when there is a mismatch between the size of the baby and the size of the placenta? Could this disparity have anything to do with fetal programming?

Barker teamed up with a Finnish researcher who had access to a rich trove of data about births in Helsinki from 1934 to 1944. The Finnish health-care system provides free medical care and a standard battery of tests for all pregnant women. The recorded information included placental weights and childhood living conditions. They found that adults who developed hypertension were more likely to have had an underweight placenta and were more likely born into poor living conditions, which suggest poor nutrition.

Based on this study, it seemed possible that maternal undernutrition was a factor in poor placental growth, since this organ gets the food it needs directly from the mother's body. Poor placental growth and function led to poor growth of the baby before birth. In turn, poor prenatal growth elevated an individual's risk of developing hypertension. Researchers are beginning to connect the dots, but we are still a long way from understanding the winding path that leads from suboptimal conditions in the womb to a heightened risk of certain chronic diseases that appear during adulthood.

OTHER FACTORS

The work of Barker and his colleagues suggested that suboptimal growth before birth and during early childhood perpetrated fetal programming. While much of their data underscored poor maternal nutrition as the trigger, many researchers questioned whether this was the only cause. They began to investigate other factors that might compromise a baby's growth during this critical period and could also have a detrimental effect on their health as adults.

An obvious place to start was the pregnancy complications that could have a negative impact on growth before birth, and consequently, on birth weight (see chapter 5). Much of the research has focused on preterm birth. A baby born before thirty-seven weeks may be small but appropriately grown for his or her gestational age at birth, or instead, may be lagging on the growth curve, a condition called intrauterine growth restriction, or IUGR. Studies show that among infants born prematurely, babies with a low birth weight (less than 5.5 pounds) or whose prenatal growth was restricted are the most susceptible to cardiovascular disease and other signs of fetal programming later in life. In other words, preterm birth accompanied by low birth weight and growth restriction elevates the future risk for some adult diseases.

Given that poor placental growth and function are implicated in fetal programming, it is logical that pregnancy complications linked to placental defects can have the same effects. Shallow attachment of the placenta to the uterus is believed to be a major cause of early-onset preeclampsia, which triggers signs of maternal vascular damage, including dangerously high blood pressure. Preeclampsia can also have negative effects on the baby, resulting in IUGR. Or shallow placentation and IUGR can occur without any maternal signs. In both cases, poorly grown babies have a heightened risk of cardiovascular disease, obesity, and type 2 diabetes.

If pregnancy complications that cause poor fetal growth can trigger fetal programming, what about environmental factors that do the same thing? Certain chemical exposures, such as maternal smoking, are associated with low birth weight. Do they also push the button on fetal programming? Researchers are beginning to suspect that this is the case. Knowing that these types of exposures could have lifelong negative effects on babies and children is another good reason for avoiding them whenever possible.

Besides these factors, there is mounting evidence that the mother's experience of stress may impact a baby's adult health. It's a given that anxieties and worries come with modern life, but how they are managed can make a difference. If a pregnant woman is dealing with stress that impacts her health — that, for instance, sends her blood pressure soaring — there can be repercussions for her unborn child.

Researchers estimate the prenatal exposure of babies to stress is widespread, affecting 10 to 35 percent of children worldwide. Studies show that regardless of its source — from the death of a loved one to a more impersonal but nonetheless traumatizing event — stress can have a profound impact.

Danish investigators used their detailed national-health databases to investigate the effects of a close relative's death during the twelve months prior to the birth of a male baby. This design enabled them to relate one of life's biggest stresses to the health of that child years later, when he underwent a medical exam prior to military service. They found that these young men were more frequently overweight or obese compared to other recruits, a result that suggests fetal programming.

Another group studied women who were pregnant and living or working close to the World Trade Center on September 11, 2001. These mothers had a high incidence of posttraumatic stress disorder after the terrorist attacks. The data revealed that their babies had reduced head circumferences at birth, which could negatively impact neurocognitive development farther down the road.

If these data are more broadly applicable, it would suggest lesser but discernible effects on the adult health of babies born to women who experience everyday stresses rather than a disaster, whether personal or national. We don't yet know if that is the case. However, we've known for decades that too much stress is dangerous for any person. Because there are many years between the instigating event (stresses that are a normal part of life for most pregnant woman) and the outcome (fetal programming of adult chronic diseases), it will take time to prove or disprove an association.

Before definitive results are in, trying to limit your exposure to stress when possible makes sense. Sometimes you *do* have a choice — whether it's turning off the news, not taking on another big project at work, or getting realistic about your to-do list. It's important to look for ways to manage the tensions of modern life during pregnancy through self-care, including adequate sleep and rest, good nutrition, and exercise. When necessary, ask for more support at work or at home to handle a situation that may be stressing you (and your baby) out. If you practice yoga or meditation or have other routines that can help relieve stress, keep them up — your baby will benefit as well. (Exercise is an excellent stress-buster for many; it's discussed later in this chapter.)

WHY DOES FETAL PROGRAMMING HAPPEN?

We still don't understand all the mechanisms behind fetal programming, though scientists all over the world have now confirmed the Barker hypothesis: a baby's growth trajectory before birth can leave a lasting imprint on adult health, including the likelihood of developing certain diseases.

Perhaps the best guess as to why this happens is that the causes are epigenetic — in other words, they don't involve changes in the DNA code. The epigenetic "traffic lights" on DNA, which we discussed in chapter 1, determine whether particular genes are active or silent. Fetal programming may disrupt these signals, changing gene expression in unwanted ways.

One way to understand how the epigenome works is to consider the special case of identical twins. They have the same DNA, making them the human equivalent of photocopies. They share the same environment before and often after birth, during early childhood. Despite this fact, identical twins can differ in certain characteristics, such as disease susceptibility. Why does one develop rheumatoid arthritis, for instance, and the other doesn't? The explanation likely involves the epigenome, those traffic lights on DNA. We know that the pattern of these signals can be altered by many factors, such as stress, diet, and exposure to environmental chemicals. When the twins grow up and go their separate ways, their different life experiences, including healthy or not-so-healthy lifestyles, might manifest as changes at the epigenetic level. The DNA of the twin who develops the disease could acquire green lights that activate genes implicated in autoimmune disorders, while the DNA of the unaffected twin retains the normal pattern of red lights, which keep these genes in the off position. As in most things, personal choice matters. In photographs of identical twins, it's easy to pick out the one who smoked cigarettes.

Evidence suggests that fetal programming involves these epigenetic traffic lights as well. One study focused on adults who were conceived just as Germany imposed an embargo on food shipments to the western Netherlands during World War II. The sudden food shortages that ensued caused a brief period of famine known as the Dutch Hunger Winter of 1944–45. Pregnant women were starving, and the babies they were carrying became malnourished. Subsequent studies showed that unborn children in the early stages of development at that time were the most susceptible to fetal programming. As adults they had high rates of the telltale diseases.

Six decades later, scientists analyzed the epigenomes of the original subjects. As compared to their siblings who were born after the Hunger Winter, the in utero victims of famine had an abnormal distribution of signals on an important growth-promoting gene. The consequences could include dialing down its activity, one of many possible reasons that these babies grew poorly before birth. Nutritional deprivation, it seems, can permanently alter a baby's epigenome, impairing fetal growth and, eventually, adult health.

Beyond discernible effects on the developing baby, some investigators, including Dr. Kent Thornburg (who collaborated with David Barker), think that the placenta is the "center of the fetal programming universe." Placental development is sensitive to the mother's levels of nutrition, a key driver of fetal programming that could target the traffic signals on the DNA of its cells. These changes might compromise the organ's functions, leaving a permanent imprint on the baby. Research in this area is a hot topic, and many scientists are trying to prove (or disprove) hypotheses related to the placenta's role.

OPTIMIZING YOUR BABY'S ADULT HEALTH

By now you may be concerned about the possibility of a vicious cycle: Fetal programming is a risk factor for developing obesity and type 2 diabetes in adulthood. Furthermore, a pregnant woman with these conditions can pass a heightened risk of the same health problems (and other chronic diseases) to her baby. And so the cycle continues.

Or does it? After all, this phenomenon is also known as the developmental origins of adult health and disease — so let's focus on health rather than disease. There is a great deal that you can do during pregnancy to mitigate the risks of fetal programming. It's even better to start before you become pregnant, but as the saying goes, there's no time like the present.

Take Charge with Nutrition

The Barker hypothesis and related theories were largely built on data about maternal undernutrition, but we now know that maternal overnutrition may also trigger fetal programming. Apply that principle to your own pregnancy: practice good nutrition!

You already know that a healthy diet during pregnancy (and before) is beneficial to the development of your baby and placenta. Now, it turns out, good nutrition contributes to the path your child's health may take over a lifetime. That's added motivation to eat well!

We simply don't collect enough detailed information about what women eat during pregnancy to study the effects of diet on the health of their babies during childhood and as adults. Due to the complexities of studying human mothers and the long-term health of their children into adulthood, much of the work on maternal nutrition and fetal programming has used animal models. But in some areas a consensus is emerging from both animal and human studies.

Eat Healthy Fats

When a mother's diet is too heavy in saturated fats (from foods such as fatty red meats and rich dairy products), it may have a negative impact on her baby's future health. Lowering total fat intake and eating a diet high in omega-3 and -6 fatty acids appears to ward off possible adverse effects of fetal programming while promoting the health of the mother.

What are some ways to boost your intake of these good fats? You can turn to seafood as an excellent source of long-chain omega-3s. (See the tips in chapter 4 on how to avoid environmental toxins, including mercury and pesticides, in your foods.) Seafoods with high levels of these fats include these:

* Salmon
* Sardines
* Trout
* Sole/flounder
* Anchovies
* Mussels
* Oysters

You can also find healthy fats in olives and olive oil, avocados, and most nuts, especially walnuts, almonds, pecans, pistachios, hazelnuts, and macadamia nuts. Seeds such as flax, chia, and pumpkin are also a good source.

Enjoy a Variety of Foods, Especially These

The most satisfying way to eat is to enjoy a variety of good foods every day. That's also how you'll get the widest range of macro- and micro-nutrients. In addition to adding healthy fats to your diet, include lean protein (from fish, eggs, meat, and dairy), fiber (from vegetables and fruits), and carbohydrates (from whole grains and starchy vegetables). Dark leafy greens, including spinach and kale, are loaded with important nutrients such as folate.

Take a Prenatal Vitamin Supplement

You can get plenty of nutrients for yourself and your baby through a healthy diet, but as discussed in chapter 2, there are good reasons to add a prenatal vitamin supplement — and staving off the effects of adverse fetal programming is one of them. There is evidence that certain vitamins can aid in prevention of obesity. In humans, low circulating levels of B12 are associated with a higher maternal BMI. In animal studies, adding folate (vitamin B9) and vitamin B12 to the diets of obese mothers can prevent this condition from developing in their offspring. Experts recommend taking a prenatal vitamin that includes these supplements to guard against neurological defects in a developing baby.

Take Charge with Exercise

Good nutrition and regular exercise complement each other, sometimes making it hard to determine the effects of one versus the other on pregnancy. Both can help control blood glucose levels, ward off hypertension, and maintain a healthy weight. (And inadequate amounts of exercise, of course, can do just the opposite.) Nutrition will give you fuel for exercise, but vigorous physical activity makes some pregnant women ravenously hungry, ready to eat whatever is on hand, nutritious or not — all the more reason to carry some healthy snacks with you, along with plenty of water. As you'll see, staying active is worth the effort. Many experts think that exercising during pregnancy is an effective way to interrupt the detrimental aspects of fetal programming. How might this work? Exercise increases blood flow and oxygen delivery to your organs *and* the placenta, promoting its development. It takes a well-grown, healthy placenta to produce a well-grown, healthy baby. The amount and timing of exercise seem to be particularly important variables in this equation.

* Women who exercise before conception and continue a moderate program of physical activity during pregnancy have placentas with increased surface areas, which, as a result, function better.
* The same is true for exercising mothers (active before conception) who dial back workouts during the last half of pregnancy.

In fact, this group had the most benefit in terms of advantageous placental structure.
* The placentas of women who begin exercising in early pregnancy and continue until the end also show many of these beneficial changes.
* Conversely, increasing exercise during the last half relative to the first half of pregnancy *decreases* placental growth.

In general, the impact of exercise on the growth of the baby parallels that of the placenta. Women who exercised during the first half of pregnancy and reduced their physical activity during the second half had babies that weighed 11.5 ounces more and were three-quarters of an inch longer than those of women in other groups. Women who increased their exercise level during the last half of pregnancy had babies with decreased birth weights.

A mother's health also benefits from exercise. A twelve-week supervised program of low-to-moderate strength training initiated at the end of the second trimester lessened the risk of lower back pain. There is also evidence that women who continue exercising during pregnancy stick to their routines after they give birth. By the time they reach menopause, they have a lower cardiovascular risk profile.

Physical activity also has a positive impact on birth outcomes. A common worry is that exercise, particularly the kind that includes a lot of up-and-down motion, could move the baby lower in the pelvis, heightening the risk of preterm birth. But for women with normal pregnancies and a low risk profile, this has never been proven. In fact, there appears to be no discernible impact on the length of pregnancy. Instead, exercise has a positive impact on birth, lowering the rate of cesarean deliveries and increasing the rate of vaginal births.

How often should you exercise, and for how long? That depends on your own health and how your pregnancy is progressing. As a general guideline, the US Department of Health and Human Services' Physical Activity Guidelines for Americans suggests that if you're not active already, aim for a weekly minimum of 150 minutes of moderate aerobic activity, spread throughout the week (for example, five half-hour workouts). If you're on the other end of the spectrum and you regularly do vigorous exercise or consider yourself highly active, you should remain active throughout your pregnancy and during the postpartum period, but talk to your health care provider about when to scale back, given the link between placental health and exercise during the last half of pregnancy.

The American College of Obstetricians and Gynecologists (ACOG) takes the official position that exercise benefits most healthy women, and that those with no pregnancy complications "should be encouraged to engage in aerobic and strength-conditioning exercises before, during, and after pregnancy." ACOG does, however, make very specific recommendations about the medical problems and pregnancy complications that make aerobic exercise during pregnancy inadvisable. They are listed here:

Absolute Contraindications to Aerobic Exercise During Pregnancy

* Significant heart disease
* Significant lung disease
* Incompetent cervix or following a cerclage procedure
* Carrying multiple babies or any other risk for preterm birth
* Persistent bleeding
* Placenta previa
* Premature labor during the current pregnancy
* Ruptured membranes
* Preeclampsia or pregnancy-induced hypertension
* Severe anemia

Although ACOG does not spell this out, a factor to consider is whether you have had a problem such as a preterm birth in a prior pregnancy. If so, you may want to play it safe and reduce your exercise regimen. As we discussed in chapter 5, a history of preterm birth puts you at higher risk for a recurrence.

ACOG also specifies medical conditions that make physical activity relatively risky:

Relative Contraindications to Aerobic Exercise During Pregnancy

* Anemia
* Maternal heart arrhythmia that has not been thoroughly evaluated
* Chronic bronchitis
* Heavy smoker
* Poorly controlled type 1 diabetes
* Poorly controlled hypertension
* Poorly controlled seizure disorder
* Poorly controlled hyperthyroidism
* Extreme morbid obesity (BMI over 40; BMI over 35 when experiencing obesity-related health conditions such as diabetes and hypertension)
* Extreme underweight (BMI less than 12)
* History of extremely sedentary lifestyle
* Intrauterine growth restriction in current pregnancy

Many chronic medical conditions are on this list. Fetal growth restriction — when the baby falls off the growth curves, progressing more slowly than expected — is another indication that it's time to take it easy, as the added stress of exercise could exacerbate the problem.

ACOG also suggests physical activities that are generally safe for women to initiate or continue during pregnancy. Consult with your obstetric care provider to double-check that your specific workout routine is advisable, especially if you have one or more of the relative contraindications listed above.

These non-weight-bearing exercises are commonly recommended:

* Swimming
* Cycling on a stationary bike

Weight-bearing activities in which you work against gravity are also encouraged. Choose exercise that does not make you feel light-headed, a potential sign of hypotension (abnormally low blood pressure) or hypertension:

* Low-impact aerobics
* Walking
* Modified yoga positions
* Modified Pilates or bar exercises

Other forms of exercise are recommended if a woman engaged in them before pregnancy. Make sure, however, that the movements do not throw you off balance — your center of gravity is changing:

* Jogging
* Running
* Racket sports
* Strength training

Any of the activities listed here should be discontinued if one or more of the following symptoms appear:

* Chest pain
* Dizziness
* Headache
* Trouble breathing (before or during exercise)
* Muscle weakness
* Swelling or calf pain
* Abdominal pain

The following warning signs are evidence of a pregnancy problem. If any of them appear, stop exercising and contact your health care provider:

* Vaginal bleeding
* Painful contractions
* Leakage of amniotic fluid

ACOG also recommends that certain forms of physical exercises be avoided during pregnancy. They include "hot" yoga or Pilates done at elevated temperatures. This advisory comes as a result of studies showing that women who are exposed to external heat sources such as hot tubs, saunas, or electric blankets have a higher risk of giving birth to a baby with a major malformation. Contact sports such as basketball, soccer, boxing, and ice hockey are not advised, nor are activities that involve a risk of falling: skateboarding, skiing, surfing, off-road cycling, gymnastics, or horseback riding. Subjecting your pregnant self to extreme changes in air pressure, such as those encountered in scuba or skydiving, is also not a good idea.

As is the case with so many aspects of pregnancy, additional research needs to be done before it will be possible to provide the best exercise regimen for every pregnant woman. We have to learn a great deal more before we can optimize the duration and intensity of exercise for each trimester of an individual pregnancy. But for healthy women with normal pregnancies, commencing or maintaining regular physical activity is one way to interrupt the effects of adverse fetal programming, reducing the likelihood that your baby will eventually develop a chronic medical condition during adulthood.

EXERCISE DURING PREGNANCY: A WORKOUT FOR MIND AND BODY

One way to handle the everyday tensions of daily life — before, during, and after pregnancy — is to incorporate exercise into your life. Besides its many physical benefits, exercise has been shown to reduce depression, anxiety, and overall stress levels. It has also been linked to improved sleep and cognitive function. And then there's the link between stress and fetal programming. All in all, the argument for regular exercise is a strong one.

If you are newly pregnant (or are trying to conceive) and you already have a physical activity you enjoy, talk to your health care provider about how to modify your routine as pregnancy progresses. Some women find that familiar workouts such as swimming, cycling, running, or brisk walking become trickier or feel different when they are pregnant. The hormone relaxin (which does exactly what its name implies — it relaxes ligaments in preparation for childbirth) may affect your physical stability, and unfamiliar muscle aches and pains may arise. Or you may just tire more easily. As the weeks and months go by, pay attention to how you feel, and continue to adjust your routine.

Though exercise generally offers psychological and physical benefits, some research about exercise during pregnancy indicates possible negative effects. For example, there is evidence that physical activity reduces rather than promotes the growth of the baby in utero. Women who participated in a cycling program from twenty weeks of pregnancy to term had babies who weighed, on average, five ounces less at birth than those of mothers who did not exercise. This finding is consistent with studies showing that women who increased their level of exercise in the second half of pregnancy had smaller placentas. (See page 184.)

Once again, the take-home message is that every pregnancy is unique. Whether physical activity will benefit you and your unborn baby falls into the realm of personalized medicine. Get your own "prescription" for exercise from your health care provider, who is familiar with your physical condition and health history. This person is in the best position to help you make informed choices.

Take Charge with Breastfeeding

Evolution has devised the perfect meal plan for your baby.

The composition of breast milk makes it particularly well suited to human babies. It has just the right formulation of proteins, fats, and carbohydrates to support optimal growth after birth. It's also one of life's great experiences to feed a "hangry" baby who awakens mad with rage because of an empty stomach and who is also terrified that food will never appear again! Few interactions will ever again be so mutually satisfying for mother and child. For that one brief moment you can give your baby everything he or she needs.

Over the past several decades, societal attitudes about breastfeeding have evolved, along with our understanding of its many benefits. When I returned to work after maternity leave, no accommodations were made for women who wanted to continue providing their babies with a steady diet of breast milk. I was lucky to have my own office, so I could lock the door and pump whenever I needed to. Others were not so fortunate. They were relegated to the restroom stalls, not the best place in the world to prepare your baby's next meal.

Though some women may lack privacy on the job (we are in the era of the "open space" office plan), things have changed at many workplaces. At my university, there are specially designated lactation rooms in almost every building on campus; they are outfitted with comfortable furniture and repurposed small refrigerators (originally built for cooling white wine rather than mother's milk — this is San Francisco). Still, pumping breast milk is an investment of time and effort. But if you plan to breastfeed and want to continue after you return to work, it's worth it.

Besides the nutritional benefits, strong evidence suggests an additional advantage to breastfeeding: it may moderate the effects of adverse fetal programming. Dr. Dwight Chapin, the physician who promoted better living conditions for orphans, suggested that what an infant eats influences his or her metabolism later in life. David Barker and Clive Osmond thought so too; they cited Chapin's ideas in their seminal 1986 paper on fetal programming. They also referenced other studies contemporary with their own, which had turned up evidence that this was true. Women who, as babies, were breastfed during the first five months of life had lower serum cholesterol levels than those who were not. A similar trend was observed among men.

Since then, much of the work in this area has focused on obesity — an outcome that's easy to measure in societies where this condition has reached epidemic proportions among children. One group studied mothers who experienced excessive weight gain during pregnancy. Breastfeeding decreased their babies' risk of subsequently developing obesity. Another study showed protective effects of breastfeeding against childhood obesity when the mother had diabetes before she became pregnant.

Researchers continue to delve into the links between breastfeeding and fetal programming. In addition, data are accumulating in favor of breastfeeding as a general preventive measure for childhood obesity. Some of the statistics are impressive. A retrospective analysis of data from several studies found that breastfeeding for only one month resulted in a 4 percent decrease in risk for obesity. Re-analysis of data from more studies showed that breastfeeding for longer amounts of time had an even greater beneficial effect, reducing by 13 percent the incidence of children becoming overweight or obese.

Breast milk is a sort of miracle drug. It gets a baby off to the right start, and we're still learning how it can impact adult health. Also, breastfeeding has health benefits for the mother, substantially reducing the risk of breast and possibly ovarian cancer. But, if for some reason you aren't able to breastfeed, don't worry. If you had a normal pregnancy and a well-grown baby, your child is already on the right path. With continued good nutrition, no matter the source, that course should continue.

Take Charge by Helping Your Baby Grow the Right Way

Many studies, including those done by David Barker and his colleagues, suggest that the way a baby grows after birth has a lot to do with whether fetal programming will manifest in adulthood. Understanding a concept called "adiposity rebound" (*adiposity* means "fat") is key to recognizing the difference between a growth trajectory that is protective and one that confers heightened risk. Normally, a child's BMI decreases from age two to six, before increasing again — that's the adiposity rebound. Or, to put it another way, from age two to six, a child usually loses his or her "baby fat."

But what happens to the adult health of children whose BMIs don't follow this pattern? In a study of a large cohort of men and women who were born at Helsinki University Hospital, Barker, Johan Eriksson, and their collaborators showed that the risk of developing type 2 diabetes was related to birth weight and the age at which a child's BMI rebounded. The highest risk was conveyed by a growth pattern in which a child:

1. had a low birth weight;
2. experienced an early onset of the rebound (for example, four years of age instead of six); and
3. sustained an accelerated increase in BMI thereafter.

Investigators studying another large group of subjects in India also found this relationship in their cohort, meaning that the results are probably true for people from many different genetic backgrounds.

An interesting aspect of this work was that obesity was not a factor. The heightened risk of type 2 diabetes was related simply to the growth trajectory. Certainly, BMI is an independent risk factor for developing type 2 diabetes, but in such cases individuals tend to have high birth weights, a high score on the ponderal index (the ratio of height to weight), and some history of childhood obesity. A mismatch between body size at birth and in childhood — for example, a low birth weight and an early adiposity rebound — is also associated with the development of coronary heart disease, in which the blood vessels that supply the heart become choked with plaque.

How does all this relate to the growth of your child after birth? Here's the bottom line: growth trajectory is much more important than typically imagined. When a baby is born with a low birth weight, it seems only natural to encourage "catch-up growth," a rapid weight gain during the first few months of life, which brings an infant closer to the normal range. Some of that is beneficial — it's the usual plan for premature babies or those who have grown poorly before being born. A good spurt of catch-up growth is, for example, predictive of neurodevelopmental outcomes better than those of babies who grow poorly after birth.

However, it's possible to overdo it. The data suggest that, after the age of two, a normal growth trajectory will ward off the possible effects of adverse fetal programming. For both low- and high-birth weight babies, aiming for the onset of an adiposity rebound at around age six, with slow and steady increases in BMI throughout childhood, can set the stage for a lifetime of good health.

Take Charge by Protecting Your Baby's Microbiome

For years scientists ignored the role of the microbiome, the trillions of microbes such as bacteria that cohabit our bodies, in human health. These organisms, residing in locations such as the gut, were once thought to be mere passengers that accompanied us on our journey through life. Over the past several years, however, this passive view has changed dramatically. We now know that the microbiome plays important roles in keeping us healthy and is involved in a wide range of diseases such as GI disorders. (The gut alone, it is estimated, hosts over a hundred trillion microorganisms from more than a thousand different species!)

It was a big surprise to scientists when they discovered that some ulcers were caused by a bacterium (*Helicobacter pylori*) and that these lesions, if left untreated, could turn into stomach (gastric) cancer. A microbe caused a substantial number of ulcers and even cancer! Almost overnight it became standard medical practice to test ulcer patients to determine if they had been colonized by this bacterium. Those diagnosed with the condition were treated with antibiotics, which cured them.

The flip side can also be true — when we kill off the "good" microbes or don't help them thrive, we raise our risk of getting certain illnesses and diseases. A whole industry has sprung up around the production and marketing of probiotics — various products containing live bacteria and yeast that are meant to encourage the growth of healthy microbes. The idea is that probiotics will counteract the *anti*biotic nature of our modern lifestyles. Whether they actually do this or not is a complicated and difficult question to answer.

Research into the microbiome (usually concerning adult intestinal health) is popular, regularly covered in the mainstream media. But you probably haven't seen as much on the role of the microbiome in the growth of a newborn baby through infancy and into early childhood. Some of this work is at the forefront of research into ways to dampen the effects of fetal programming. As a result, most of the data come from animal rather than human studies. Nevertheless, the results are intriguing.

A child's gut microbiome develops during the first three years of life, as he or she begins to eat a wider range of foods. Healthy growth is associated with successive colonizations of different bacteria, which come and go in predictable ways. If you are already a mother or if you have experience with the digestive tracts of infants and young children, you will be very familiar with the GI symptoms that accompany these fluxes — the onset of new tummy aches and lots of expulsions from both ends! They are the gut's version of the normal growing pains experienced by a healthy child.

Investigators have examined the course of bacterial colonizations in babies who fail to grow normally. Some of the initial studies focused on malnourished children. Researchers found that their gut microbiomes were immature. When this mix of microbes was transferred to mice, the recipients experienced poor growth. Thus, malnutrition produces changes in the microbiome that further impair normal development, a double whammy for the most vulnerable children in the world.

Conversely, what promotes the establishment of a healthy microbiome that enables normal growth of a baby? Some researchers think that breast milk might be playing a role. It contains special sugars that feed not only the baby but also the bacteria that reside in the GI tract. As a result, breastfeeding may help establish the types of bacteria that promote good growth and may prevent bacteria associated with poor growth from gaining a foothold. Some investigators also think that breastfeeding promotes, in the child, a microbiome that makes it hard for microbes associated with obesity to take up residence.

Currently, studies aimed at improving the growth of low-birth-weight babies by targeting the microbiome are far from conclusive. But this active area of investigation is well worth watching because transformative therapies could emerge. In the meantime, helping your baby establish a healthy microbiome through breastfeeding is one way to establish a normal growth trajectory.

IMPROVING MATERNAL NUTRITION
GLOBALLY — AND LOCALLY

If you're reading this book, you're clearly interested in doing all that you can to nourish your baby in the womb, and no doubt you'll do your best to instill in your child healthy eating habits for years to come. But in some parts of the world, parents who wish to do the same thing face huge obstacles.

Unfortunately, in resource-poor regions, being born small is the major cause of neonatal and early childhood death. And those who survive have a significantly increased risk of suffering the harmful effects of adverse fetal programming, further compromising the adult health of the most vulnerable populations on the planet. In low- and middle-income countries, a staggering number of babies are born every year whose growth before birth is far below optimal. In these settings, infectious disease is a major culprit in low birth weight. Malaria, in particular, causes anemia in the mother and also damages the placenta, impairing its ability to transport oxygen and nutrients to the baby. These conditions lower a baby's birth weight.

But low birth weight isn't a problem only on the other side of the world. In your own community, there are pregnant women who are not able to adequately nourish themselves — and their unborn children — and the negative effects of fetal programming will be hard to escape or reverse.

At the time of his death in 2013, David Barker was trying to improve maternal nutrition in Portland, Oregon, where he was working with colleagues at Oregon Health Sciences University to trace the winding road that leads from fetal programming to health problems in adulthood. Although his theories had gained wide acceptance, he was still being criticized — this time for not devoting 100 percent of his time to figuring out the mechanisms behind fetal programming. But he had unshakable faith in his own data and the long-term societal impact of his findings. He was inspired to take action in his own community, an example well worth following.

7

Birth

Every pregnancy has its own story, as unique as the child who is born. It's also true that each birth is unique, beginning with labor and ending with the delivery of the baby and the placenta. In this final chapter, we'll cover the aspects of birth that are unique to humans. In turn, this knowledge will help you make informed decisions about how you want to approach labor and delivery. But some of the best information you'll gather won't come from a book, a website, or other such resources. Instead, the true experts take over here — women who have given birth.

LEARNING FROM BIRTH STORIES

When you become pregnant, you'll be showered with congratulations and peppered with advice by veteran moms, much of it having to do with pregnancy itself and the care of a newborn. You'll get opinions on how to handle morning sickness, where to order the most comfortable but stylish maternity clothing, which car seat or breast pump to purchase, how many clean onesies you should stockpile, what to carry in the diaper bag, and favorite remedies for colic — all of it based on personal knowledge. One of the most important things that women can do for one another is share their experiences of birth. Yet many mothers go quiet on the topic. Perhaps they think they're doing you a favor, sparing you the "gory details," or they may not want to discuss the topic for personal reasons.

One of the best ways for a first-time mother to gain a better understanding of labor and delivery is to break through this barrier and ask women with whom you have a close relationship — a mother, sister, or best friend — to give you an honest and detailed account of their own experience. This advice also holds for women who have already given birth, since it's possible to have a very different experience the next time around.

Choose women whose approach to most things medical closely mirrors your own. If your sister selected her dentist based on his willingness to give her nitrous oxide sedation for a tooth cleaning, hers is probably not the story you will want to seek out. Likewise, if your best friend walked around for two days with a broken leg (as mine did) because she has an iron will and a pain threshold to match, she may not be your go-to person either.

Also, collect stories from women who have had a range of birth experiences. Some accounts will be easier to solicit than others. Women who went into labor, quickly progressed through the stages, pushed three times, and delivered the baby are likely to willingly share their experiences. Mothers who had a long labor, a twenty-four-hour-plus marathon that ended with a cesarean section, may be less forthcoming. Pain control is an important issue to discuss. Did they talk to their obstetric care provider beforehand and choose a method? What actually happened? What part of the process would they change if they had another baby?

At the same time, the person who will be at your side to support you through labor and delivery — your partner, close friend, or relative — should go on a fact-finding mission too. Your support person needs to be prepared. When you are in labor, you don't want to be coaching your labor coach!

Share the stories you collect and work them into your own hypothetical narrative, your own birth plan. Being on the same page with your support person is vital. It's also important to acknowledge that this plan may go out the window. Birth can be like riding a roller coaster in the dark: you have to be ready for unexpected twists and turns. Considering alternative scenarios is itself an important part of a birth plan.

Another way to prepare for birth is to arm yourself with knowledge about the biology. It's true that there is a lot we don't understand about the whole process, starting with what initiates labor, but that doesn't mean it's entirely mysterious. There are important facts to consider, starting with the mechanics.

HUMAN BIRTH: A TIGHT FIT

In most species, the placenta forms a superficial attachment to the uterus (see chapter 2). Therefore, delivery of offspring and the placental separation that follows are relatively simple. But in humans, the unique relationship between placenta and uterus makes the delivery of our babies a one-of-a-kind process. The placenta buries itself deep inside the uterine wall, where it forms elaborate connections with maternal blood vessels. After labor and delivery, its roots must be extracted. At the same time the arteries must quickly clamp shut to stop a rush of your blood into the now empty cavity of your uterus. If this tightly orchestrated process does not go smoothly, significant bleeding can occur. That's one reason why it's a good idea to have the birth of your baby attended by an obstetric care provider who knows how to handle this situation. Also, a hospital setting adds another measure of safety. If there's a bleeding problem, the necessary medical equipment and staff are right there to address the problem.

But the placenta isn't the only issue that makes birth more complicated. As humans went from walking on all fours to standing upright, our pelvises narrowed to hold in our organs. Around the same time, human brains grew bigger, and the surrounding bones of the skull enlarged to accommodate their greater volume. As a result, the head is the largest part of the baby's body. This created a new difficulty, the

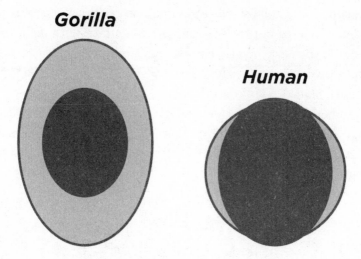

The head of a baby gorilla (*solid dark circle to the left*) easily fits through the mother's pelvic outlet, which is shaded a lighter color. For humans, it's a tight squeeze (*solid dark circle to the right*).

geometry problem shown in the illustration above. During delivery, the baby must travel through the birth canal, a passageway in the pelvis where the right and left hip bones join in the middle. Most other animal babies, even some of our closest nonhuman primate cousins (such as the gorilla shown in the illustration), have heads that are much smaller than the outlet of the birth canal, allowing easy passage during delivery. In contrast, our large heads have reached the upper limit of what can reasonably pass through this opening. Sometimes it's a very tight fit. The reason why a newborn baby's head may look a bit squashed? It could have been.

And then there are your baby's shoulders. Try this: rotate your shoulders inward, as far as possible, toward the middle of your chest. There is only so much you can do to narrow the breadth of your shoulders. Likewise, there is only so much that your obstetrician or midwife can do to compress your baby's shoulders to ease passage through the birth canal, which usually takes a couple of twisty turns to accomplish.

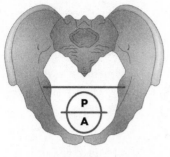 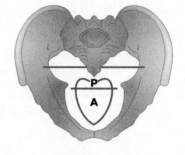

Female   **Male**

The female pelvis (*left*) has a round-shaped opening, a favorable passageway for the birth of a baby because the head is more likely to fit. The male pelvis (*right*) has a heart-shaped opening. Sometimes, however, the geometry of a woman's pelvis resembles that of a man, which can impede birth. P stands for *posterior* (toward the back); A stands for *anterior* (toward the front).

Another variable is the shape of your pelvis. As with other body parts, there is quite an array of variations. During puberty, under the influence of sex hormones, the pelvis takes on a different shape in boys and girls. In males, the pelvis is heart-shaped. In females, the evolutionary pressure exerted by the need for a baby to pass through the birth canal has rounded out this space. (This distinction is used in forensics and archeology to determine whether an unidentified skeleton is from a man or woman.)

But differences between the pelvic shapes that characterize women and men exist on a continuum. Some men have a more rounded pelvis, and some women have a birth canal with a heart-like contour. Less commonly, a mother's pelvic outlet may be significantly narrowed in one direction or another. These deviations from the optimal round shape can make birth difficult.

Given that passage of human babies through the birth canal is such a tight squeeze, it's a good thing that there is some give and take on both sides. The bones of the infant's skull are not yet fully formed, so up to a point, the head can be compressed. In addition, the mother produces the hormone relaxin, which causes a loosening of the connective tissues that hold the pelvic bones together. This helps distend the walls of the canal.

All of this means that human births can be more complex than those of other species. Though the birth process usually works as it should, in some cases, there's a mismatch: the baby's head, shoulders, or both are simply too big to fit through the birth canal. Another way out has to be found: a cesarean section.

The limits of anatomy are yet another reason to have regular prenatal care. An experienced provider will do a clinical exam to determine whether your risk of having a surgical delivery is low or high. Currently, there is not enough evidence to suggest that an imaging study such as an X-ray will increase the accuracy of the prediction. In equivocal situations, the only way to find out is a "trial of labor," in other words, an attempt at vaginal birth.

PREPARING FOR YOUR BABY'S ARRIVAL

As pregnancy advances, your baby comes into sharper focus. In an amazing process that began with a fertilized egg, he or she has been transformed from a primitive version of a tiny human during the first eight weeks following conception to the newborn who will soon be in your arms.

If you have had multiple ultrasounds over the course of your pregnancy, you have been privy to the remarkable growth and sculpting that, by the second-trimester anatomy scan, produce a baby that looks like a human child. The third-trimester growth spurt turns the flutters that were the first sign of your baby's movements into the writhing of a contortionist who is outgrowing the cramped space offered by the uterus.

For many parents-to-be, preparing a home for a baby keeps pace with its growth, ramping up toward the end of pregnancy: the crib, the changing table, the tiny clothes, the toys. Other tasks are worth doing before your baby is born, starting with a written birth plan, a great way to organize all your thoughts about how you would like the delivery of your baby to happen. (Especially if it's early in your first pregnancy, you may still be gathering information about typical elements of a birth plan, such as pain control. There's time to research your options, and you'll also find many of these topics covered later in this chapter.)

Take Charge by Writing Your Birth Plan

A checklist of sorts, a birth plan provides assurance that you have not overlooked something that would be good to consider ahead of time. A birth plan can also serve as a communication tool — a way to make sure that you and the person (or people) who will be supporting you during labor and delivery are on the same page about your preferences.

Sharing this information with your obstetric care provider keeps both of you in the loop. Labor and delivery can sometimes unfold in unexpected ways, and decisions may need to be made quickly. Having alternatives written down can help you feel in control even when plan A becomes plan B. At the same time, the byword is flexibility. Things rarely go exactly as planned.

There are many ways to create a birth plan, and you'll find a variety of templates on the internet. Look for one with a level of detail that feels right for you, such as the one created by the March of Dimes. The short printable form captures these types of essential information:

Before delivery:

* Where you will deliver
* Names and phone numbers (of your obstetric care provider, your support person or people, and your pediatrician)
* Position(s) you prefer for labor (lying down, standing, and so on)
* Type of pain control (if any)

* Who should be told first (you, your support person, or both) if a problem arises

After the baby is born:

* Who will cut the umbilical cord?
* Will you be banking the cord blood?
* Will your baby boy be circumcised?
* Will you breastfeed?

Some women prefer a more detailed birth plan that covers different medical practices and procedures, ranging from routine to rare. You may be asked to make choices about the following details:

* *Intravenous (IV) access (a thin tube with a stopper inserted into a vein).* If needed, this port can be used for the rapid delivery of drugs or other substances. The benefits of this safety measure may outweigh the unpleasantness of a needle stick and having a small piece of tubing taped to your arm.
* *Continuous delivery of fluids.* The port can be hooked up to a bag that delivers an IV solution at a steady rate. This greatly lessens your chances of becoming dehydrated, which can slow down labor. The advantage is that you don't have to remember to drink fluids when so much else is going on or you are feeling nauseated. The downside: you are tethered to an IV pole, which goes everywhere you do.
* *Continuous fetal heart rate monitoring.* Being forced out into the world is stressful. Each uterine contraction should cause the baby's heart rate to increase. There are two common methods of continuously monitoring this rise and fall. In one, a stethoscope-type device uses ultrasound to make intermittent measurements of the fetal heart rate. In the other, an electrode is attached to the most accessible part of the baby (usually the scalp via a vaginal approach). These procedures eliminate the chance that periodic measurements might miss episodes of an abnormal

heart rate or that a busy care provider responsible for several patients might have to skip a few readings. Although this monitoring might feel intrusive or overly technical, it remains the best way for obstetric care providers to monitor your baby's status as labor unfolds.

* *Episiotomy (a cut made in the posterior wall of the vagina to facilitate birth).* This once-routine practice was often done without asking the mother. Though the American College of Obstetricians and Gynecologists now recommends against the routine use of this procedure, it is still done in rare cases to prevent severe tears. You are the one who should decide if this is acceptable to you.

Your obstetric care provider and the facility where they work may also have a standard form they use as a birth plan. It's a good idea to familiarize yourself with the information they are trying to capture. It could spare you from spending precious time to answer questions when you would rather be settling down in your room or getting an epidural.

Because a wide range of templates for birth plans are available, without too much effort you'll be able to pull together the parts you find most useful to create your own customized version. Birth is a very personal affair, and it may be worth it to you and your partner to spend time developing a unique plan that captures all the elements important to you. For example, you might intend to have a doula, a nonmedical person who serves as a coach, assist you and your support person(s) during labor. You should also plan what to do with the placenta, which has made your pregnancy possible from implantation through delivery. Good (and bad) options are discussed later in this chapter.

All the while, it's important to remember that a birth plan is just that. It's the way you *hope* to navigate labor and delivery. It's not written in stone. When circumstances dictate deviating from the course you laid out, you need to quickly adapt. Your obstetric provider will guide you in doing what is best for you and your baby.

THE BIRTH CERTIFICATE

You may want to consider filling out a draft birth certificate for your new baby *before* birth, particularly if you have decided on a name. Check with your state's department of health to confirm the information that is needed. You can also find blank templates online.

Although most of the information — date and time of birth once it occurs, the location, the parents' names, and so on — will be at your fingertips, many details are required, and you might want your support person to take the lead while you rest or tend to your new baby. Mistakes — misspellings, wrong dates — can be a headache; filing an amended version can entail significant delays. The baby's name is required. If you are having trouble choosing one, it's important to know that once your child is born there is limited time to decide — if you want to receive the birth certificate promptly.

Over the years, I have been constantly surprised by the number of times that I have had to produce my children's birth certificates. It's required for enrolling in school, applying for a driver's license or passport, getting certain types of employment, and more. Decide on a place where the document can be kept safely, and don't lose it. From experience, I know that you won't want to spend hours finding this piece of paper the night before it's needed!

Most hospitals also have you fill out an application to obtain a Social Security number for your baby, another critical form of identification that is easier to get sooner rather than later.

Take Charge by Enrolling in Childbirth Classes

Attending a well-thought-out class can be an excellent way to prepare for labor and delivery, arming you with knowledge and advice that can help you make informed choices before, during, and immediately after the birth of your baby. These classes typically take place during the third trimester, and they can vary widely in format.

Experts think that women and their partners, or other support persons, need about twelve hours of instruction to learn effective methods for birthing a child. Most childbirth educators advocate for classes spread out over several weeks, allowing women to practice the techniques as they learn them. However, depending on your schedule, you may prefer the crash-course approach, with all the information given in a single session.

If you are shopping for a class, a good resource to start with is a consultation with your obstetric care provider. The hospital where you plan to give birth may offer such classes, which may be less expensive than other options. Before selecting a class, consider the teacher's qualifications. Has she had specialized training in educating parents-to-be? Several organizations, such as Lamaze, offer an established curriculum taught by certified instructors who are required to keep up with the latest approaches for managing labor and delivery. And approaches can vary. Some teachers have a more holistic approach, while others focus primarily on medical needs. A combination of the two is often the most helpful for women and their support person(s).

What you don't want is an instructor who is more interested in relating how she managed *her* own perfect birth experience rather than preparing *you* for yours. A friend of mine who opted for one-on-one instruction got a lesson in imagining how labor was like lying on a beach, with uterine contractions washing over her body like waves upon the sand. Avoid a teacher who embraces a one-size-fits-all birth plan, especially if it's based solely on her "great" experiences.

Another thing to consider is class size. A small group setting, with no more than eight to ten couples, allows time for individual instruction. It's also important to have plenty of chances to ask questions. The other participants will likely ask about issues that you and your partner may not have thought of, or are reluctant to discuss. Surely someone will ask whether it's safe to have sex as birth approaches. (The answer is yes, unless your clinician tells you otherwise.) If some of the participants have already had a birth experience, they can contribute real-world information about delivery.

Avoid crowded classes in which participants are lectured on how birth happens at a particular hospital and are offered a one-size-fits-all birth plan. This type of instruction can make it seem that you have few choices and may make you feel powerless.

There's a basic set of information that all childbirth classes should provide:

* The changes in your body that enable childbirth, such as shortening and dilation of the cervix
* What to expect when you arrive at the hospital or birthing center, including interventions such as an IV or fetal monitoring
* Strategies for coping with the pain of labor, including breathing and relaxation techniques
* Options regarding pain relief measures, including epidural anesthesia, and how to determine when you need this type of support
* Common complications such as stalled labor and how they are handled
* What happens during a cesarean section

Some classes will also offer information about the care of a newborn and how breastfeeding is initiated. Additionally, you can seek out specialized classes, such as instruction on multiple births or various aspects of infant care.

A childbirth class may have an unanticipated benefit: it's a great place to meet people who are at the same stage of life — a ready-made group who will have children of the same age. It's not surprising that friendships made in childbirth class sometimes continue long after the baby is born.

CENTERING YOUR PREGNANCY

You may live in a region where a relatively new delivery approach, which joins childbirth education and prenatal care, is available. Called Centering Pregnancy, it was the brainchild of a nurse-midwife, Sharon Schindler Rising. About twenty years ago, she found that many of her pregnant patients asked the same questions. Addressing them in a meaningful way took more time than she had in her schedule, and she was constantly running late for appointments.

Her solution? To deliver care in a small-group setting where questions and concerns could be efficiently addressed, leaving time for discussion and other activities such as childbirth education. Over time this "centering" concept evolved and now has a fairly standard format. Expectant women who are due at about the same time meet, in two-hour blocks, with an obstetric care provider about ten times over the course of pregnancy. Partners are welcome. Sometimes the meetings continue for a few months after birth, with the focus shifted to infant care.

The sessions are designed to empower the participants and to turn a group of individuals into a community. The women take turns leading the group. They are encouraged to take charge of the discussions, posing questions and answering them rather than listening passively as a health care provider lectures. They also do portions of their own prenatal care. The women take turns weighing themselves, measuring their blood pressure, and recording the results in their charts. Each participant also has a short one-on-one visit with a health care provider to check the developing baby's heartbeat and do ultrasound scans if indicated. This personal meeting time gives women a chance to ask questions privately.

As yet, no definitive studies show whether group prenatal care results in better birth outcomes. But small-scale evaluations of Centering Pregnancy groups have turned up some encouraging findings. Some participants rate the group experience as superior to traditional ways of delivering prenatal care and teaching childbirth education classes. A theme that emerged from a study that recruited pregnant women in the military and wives of military members can be summed up by a single quote: "I was not alone." More surprising, some studies show that the group approach significantly reduces certain pregnancy complications for high-risk groups: low-income, Black, Hispanic, and adolescent participants. Rates of premature birth were lower, and birth weights were higher.

If you are interested in this approach, check with your health care provider. Centering Pregnancy often works best when the participants come from similar backgrounds and are also similar in age. In some areas these programs are still hard to find, but the movement is spreading.

Take Charge by Taking the Hospital Tour

Hospitals and other facilities offer tours of their labor and delivery suites, sometimes as part of a childbirth class. It's important to take this opportunity for a dress rehearsal of your trip to the hospital, to help you figure out how long it will take you to get there from home or work.

During the tour, you will learn about the admission process and what to expect when you arrive at the labor and delivery unit. You can see the rooms where you will be before and after the baby is born. As you become familiar with the surroundings, consider how to personalize the environment and your experience. Some hospitals offer virtual tours online so you can revisit the space and, at a more leisurely pace, think about how you would like to use it.

Many hospitals provide instructions on what to bring with you and what to leave at home. You'll want to customize this packing list, but it's a good place to start — and start early, well before your due date.

You may be given a phone number to call when you think you're in labor, and it will connect you to a hospital staff person. Or you may be instructed to call your obstetric care provider. These professionals can be very helpful in determining whether the changes you observe are false alarms, thereby preventing an unnecessary trip — or they can alert you when the details you describe are signs of impending birth, requiring that you hasten your departure for the hospital.

On the tour you'll learn about the support team who will be caring for you during labor. In most hospitals, this includes an obstetric nurse, who will communicate with the midwife or doctor delivering your baby. This person or another nurse will likely spend the most time with you and be your strongest advocate during labor.

An anesthesiologist should also be part of the team, regardless of whether you plan to have an epidural or other type of pain relief.

If visiting policies are not explicitly stated, ask about them. Knowing in advance how many people are allowed to see you, and under what circumstances, can prevent misunderstandings or an uncomfortable situation in which family members or close friends are asked to leave. Note where the waiting area is and, if applicable, the hospital's policy about visits from children, such as younger siblings. Also, check on the rules about who can be present during a cesarean section. Some hospitals, but not all, allow a doula in the labor room.

As they do for arrival and check-in, hospitals have a set safety protocol for leaving. Some may even have a staff person escort you to your car, to check that your baby is correctly buckled into the infant car seat. This is something worth practicing ahead of time. Though it depends on the seat, this procedure is not always intuitive and can even be baffling. The hospital may also ask for the name of your pediatrician. It is important to choose this doctor well before your delivery.

Where Do You Park the Car?

When you are in the throes of labor, you will not want to deal with this question! If you plan to arrive in a car driven by the support person you want with you during labor, find out in advance where to park and whether you need a special permit. This information is usually covered during the hospital tour. If not, ask.

Though it seems like a tiny detail, here's why this is important: if you don't know where to park, your partner will probably pull into an emergency drop-off area, and chances are, a car can't be left there unattended. While your driver goes off to hunt down a space, you may be left on your own with hospital staff as you're admitted. Ideally — for reasons both physical and emotional — it's better to have your support person with you every step of the way, starting with helping you to register and settle in.

As for getting to the hospital, figure out in advance your quickest routes. If you go into labor on a quiet Sunday morning with no cars on the streets, you'll arrive in record time. Or it could be raining at rush hour at the start of a holiday weekend. You may be at work, minutes from the hospital, or perhaps you'll be at home and much farther away. Babies ready to be born don't consider these various scenarios — but their mothers can. You'll avoid a surprising amount of stress by sorting out these details ahead of time!

If You're Planning a Home Birth

About 1 percent of the babies born in the United States are delivered at home. Three-quarters of these births are intentional, and the rest happen by accident because there was not enough time to get to a medical facility. As rural hospitals close or stop offering obstetric care, unintentional home (or car!) births are certain to increase.

If you are thinking about a home birth, you should know that the American College of Obstetricians and Gynecologists (ACOG) advises that delivery take place in a hospital or at an accredited birth center. But you have the right to choose where you have your baby. As with so many other aspects of pregnancy, making an informed choice is key to maximizing the chances that a home birth will be safe.

For many women, this issue doesn't take a lot of thought. A home birth evokes strong feelings, one way or the other. The majority will choose delivery in a medical facility — but some women are firmly committed to giving birth at home. Frequently cited reasons include a desire for fewer medical interventions, including induction of labor and monitoring of the fetal heart rate. Others may not like the institutional atmosphere of most hospitals. For some women, the decision is based on experience. Having previously had a smooth delivery with minimal interventions in a hospital, they opt for staying home.

One problem you will encounter when trying to learn the facts about home birth is that there aren't very many available. Birth doesn't lend itself well to the usual research protocols. Randomized clinical trials are the gold standard by which the safety of a procedure is determined. Researchers assemble a study population composed of people who are as similar as possible and then randomly assign them to groups that receive different treatments. It's easy to understand why women would be reluctant to sign up for a study that will assign them randomly to a hospital or a home birth. Most mothers-to-be do not want to leave this important decision to chance.

Nevertheless, some statistics are available to consider. It may be hard to hear this, but a baby is twice as likely to die in a home birth. The risk of a newborn having a serious neurological complication, such as a seizure, increases threefold. For the mother, problems decrease with each subsequent pregnancy, if the previous births were normal and did not require significant medical interventions. A quarter to a third of first-time mothers who attempt a home birth end up being transferred to a hospital. This number falls to below 10 percent for women who have had a previous easy birth with minimal assistance, a smaller number but still a significant proportion of those who attempt home birth. In either case, the statistics suggest that another important consideration is whether rapid transport to a hospital is possible from the home. Dr. Sig-Linda Jacobson, an obstetrician and maternal-fetal medicine specialist, tells women who are considering a home birth that the outcome of obstetric emergencies outside a medical facility often depends on simple geometry: the distance between the mother and the operating room.

Women who are healthy and have sailed through pregnancy with no complications in the past are the best candidates for a home birth. But even in such cases, it is important that the baby be in the "head down" position and fully mature but not late — he or she should arrive between thirty-six and forty-one weeks of pregnancy.

ACOG warns against a home birth in the following situations:

* You are carrying multiple babies.
* Your baby is in a non-optimal (breech) position.
* You had a cesarean section for a prior birth — there's an increased risk of uterine rupture if a subsequent vaginal delivery is attempted.

The risks involved in a home birth can be reduced by having a highly educated obstetric specialist attend the delivery. The person should also be well connected with local medical-care systems in case anything goes wrong. A home birth attendant can be a midwife or nurse-midwife who is licensed by the American Midwifery Certification Board or the American College of Nurse-Midwives. Steer clear of other "licensed" midwives; in some states the recipients have undergone a registration process only, having fulfilled no training or educational requirements. Some physicians may also be qualified to perform home deliveries. All in all, choosing to have a home birth is a big decision. It should not be made lightly, and only after careful consideration of your risk factors and the qualifications of the obstetric care provider who will be involved.

WHEN YOUR BABY IS IN THE WRONG POSITION

As anyone who's been pregnant knows, babies move around a lot in the uterus, kicking, punching, stretching, and all but doing cartwheels. As pregnancy advances and growth accelerates, there's less and less wiggle room, and the dance party winds down to a yoga session as a baby's movements become constrained.

The uterus is pear-shaped, with the widest end at the top beneath the mother's ribcage and the narrowest at the bottom, near the cervix. For most of gestation the baby is upright. At the end of your pregnancy, the best way for a baby to fit into this confined space is upside-down, with the back of the head ready to enter the birth canal. This orientation, optimal for delivery given the structure of the pelvis and birth canal, is called the head-down position (or alternatively, the cephalic or vertex position).

Though most babies will assume the head-down position in time for delivery, other presentations, known as breech positions, are possible. They increase your chances of having a cesarean section. The baby's bottom may be directly adjacent to the birth canal. It can act like a stopper. Or the baby may be oriented feet first, making the initial descent through even a partially dilated cervix relatively easy, but increasing the risk of the head getting stuck later on. In that case, the umbilical cord may also emerge from the birth canal first. If the cord becomes compressed when the baby follows, this critical lifeline, which is still connected to the placenta and the mother, is cut off.

Some babies take their time getting situated in the right position. As gestation progresses, the incidence of breech presentation falls. By thirty-two weeks, about 90 percent of babies are head down, and that number increases to nearly 97 percent at term. It is likely that you will notice the accompanying shape change of your abdomen as the baby shifts lower and protrudes farther forward.

How will your obstetric care provider know if your baby is oriented correctly? An initial diagnosis is made by a physical exam. A clinician presses firmly on your abdomen to locate your baby's head and bottom. But studies have shown that this is far from an exact science. Ultrasound is used to confirm a breech presentation.

If your baby is still in the wrong position at term, your health care provider may discuss with you the possibility of having a procedure called external cephalic version. Pushing from the outside, one or two clinicians try to maneuver the unborn baby into a head-down position. It's usually done around thirty-seven weeks, when there is ample amniotic fluid to facilitate movement of the baby. In the rare case when a complication arises, delivery can be done without risking the problems that are common in premature infants.

This low-tech method actually works. A recent analysis of data from several studies showed a 60 percent decrease in breech births and a 40 percent reduction in cesarean sections. Ask your health care provider about what to expect during the process — how long it might take and what the recommended preparations are. Some women describe the procedure as uncomfortable, while others experience little to no pain. Ultimately, it could help you avoid a cesarean section.

What happens if the attempt to turn the baby fails? In the United

States, 95 percent of women whose babies are in a breech position at term have a cesarean section. Experts think that the low percentage of vaginal births among women carrying breech babies has been responsible for the substantial decrease in serious problems associated with these deliveries.

HAVING A PLANNED C-SECTION

Faced with the prospect of labor and delivery, some women elect to have a cesarean section, planned in advance, without a medical reason such as breech presentation. Often this decision involves a private conversation between a woman and her clinical care provider. As a result, it's difficult to know for sure how many C-sections are purely elective, but it is estimated to be around 10 percent of surgical deliveries. Given that one-third of all US births occur by cesarean delivery, this is not a trivial number.

If you are likely to have a normal vaginal birth — that is, if you have no extenuating circumstances such as a baby in a breech position — but prefer to have a C-section, you should think carefully. It is, after all, major surgery, involving anesthesia as well as a longer period of recovery. The abdominal incision will leave some scarring. If you are worried about the pain of labor and delivery, childbirth education classes can be helpful, teaching you coping mechanisms and ways in which a partner, relative, or friend can offer support. Remember, pain relief is available. There's no need to take an "all or nothing" approach — either entirely natural childbirth or surgery. There is a middle ground when it comes to medical interventions for pain control.

For most women without medical grounds for a C-section, the risks of the surgery outweigh the benefits, especially when it comes to future pregnancies. The uterine scar that this procedure leaves can make it difficult for another placenta to land in the proper spot, leading to a partial or complete previa (see chapter 2). Alternatively, a future placenta may decide to implant on the site of the former incision, which gives the trophoblast cells easy access to the deeper parts of the uterus from which they are normally excluded. This makes extraction of the placenta at birth even trickier than normal. Although rare, the incidence of this serious condition, called accreta spectrum disorder, is rapidly increasing. There is also an elevated risk of uterine rupture if labor is attempted in a subsequent pregnancy. The odds of these problems rise with the number of cesarean deliveries a woman has.

Also, there can be a downside for the baby. Those delivered by cesarean have higher rates of respiratory problems.

Nevertheless, there's no denying there *are* some benefits. Principal among them are lower rates of serious bleeding following birth. It's easier to make sure the entire placenta has been delivered. (Retention of small pieces in the uterine blood vessels can cause significant blood loss.) If hemorrhaging does occur, doctors have better access to the site, making stanching it a quicker process.

After weighing all the pros and cons, you may still want to have an elective cesarean delivery. Experts recommend waiting until your thirty-ninth week, to give your baby as much time as possible to mature.

PAIN RELIEF DURING LABOR

The time to think about pain relief is *before* you go into labor.

If your obstetric care provider does not bring up the subject during a prenatal visit, then you have to. This important conversation may also include an OB anesthesia specialist, who can help explain the options. Before making a decision, you need the facts. After you decide, it's important to make your preferences part of your written birth plan. If you don't, the assumption may be that you opted out of pain control when, in reality, you didn't. Or you may miss the window of opportunity for administration of the method you chose.

During the birth of our first child, I had no real idea of what to expect. I had heard that natural childbirth with no drugs was safer for the baby. Without checking the facts (extremely silly for a scientist), I blithely decided to follow my intuition — that a drug-free birth was the way to go. I wanted to do everything possible to ensure the safe delivery of our child. By the time I changed my mind and desperately wanted an epidural, I was told that it was too late. Without realizing it, the decision had been made for me.

ACOG uses clear language on the subject of pain control during childbirth: "There is no other circumstance in which it is considered acceptable for an individual to experience untreated severe pain that is amenable to safe intervention while the individual is under a physician's care ... A woman who requests epidural anesthesia ... should not be deprived of this service based on the status of her health insurance."

If you are thinking about pharmacological methods of pain relief during labor and delivery, you are not alone. In fact, it's a myth that most women choose a drug-free delivery. A survey of twenty-four hundred new mothers showed that 83 percent opted for some form of pain medication. The most popular choice, made by 67 percent of the respondents, was epidural anesthesia.

This type of anesthesia is named for the location where the medicine is placed. The drugs, usually a numbing medicine and a narcotic, are delivered to the "epidural space," the area just outside the membranes that cover your spinal cord, through a needle connected to a small piece of tubing. The injection device is connected to an infusion pump, ensuring that effective levels of medication are maintained. In a modern epidural, women are not "paralyzed" from the waist down. You will be able to walk to the bathroom with assistance. This method of pain relief is administered by a highly trained health care professional, such as an anesthesiologist. To get one you'll need to have your baby in the hospital.

Do pain-relieving drugs make labor and delivery more risky for you or your baby? Strong evidence supports the conclusion that regardless of when epidural anesthesia is initiated over the course of labor, there is no increase in surgical deliveries. Labor is prolonged, on average, by a little under fifteen minutes, with no ill effects on the baby, either before or after birth.

Other forms of pain relief are less expensive and can be given by medical staff with no specialized expertise, but these methods are also less effective. They include administration of opioids, such as morphine, which also has potential side effects. For the mother, they include nausea and vomiting. Opioids also cross the placenta and reach the baby, sometimes lowering the fetal heart rate. Inhaled nitrous oxide also falls into this category — easy to administer but generally less effective than an epidural.

The bottom line is that you have the right to choose if and how you want to control the pain of labor, and your obstetric clinician and support person should support your decision. The safe methods that are available today give you a real choice in the matter. Just because birth has happened one way for thousands of years does not mean it's the best way for you.

SERIOUSLY BAD BACTERIA: GROUP B STREP

Group B streptococcus (GBS) colonizes the vagina and rectum of about a quarter of all women, and it's a troublemaker, especially during delivery. It can travel, most frequently causing infections of the urinary tract and perhaps the uterine lining. A much bigger problem ensues if a baby picks up this microbe as it passes through the birth canal during delivery. GBS growth can explode during the first week of life, causing bacteremia (an infection in the blood), pneumonia, or sepsis (a response to infection so severe that it damages cells and organs). As a result, this bug is the leading cause of infectious disease and even death in newborns.

Fortunately, modern obstetrical care takes this potentially devastating problem into account. The CDC recommends screening of all pregnant women between the thirty-fifth and thirty-seventh weeks of pregnancy. The test involves using a swab to take a sample of secretions from the rectum and vagina, growing the bacteria they contain in a laboratory dish, and determining if they include GBS. If they do, it means you will be considered "GBS+."

This screening is important because effective strategies can be used to prevent your baby from being infected with this nasty bug during delivery. Antibiotic treatment immediately prior to delivery thwarts the transmission of this dangerous bacterium to your baby. Once you go into labor, you will be given an intravenous dose of penicillin. On-the-spot intravenous treatment is considered more effective than a traditional course of oral antibiotics. If you are allergic to penicillin, you may receive vancomycin or another antibiotic instead. The timing is important, since treating the infection too soon gives the bacterium a chance to regrow, once again endangering the baby.

LABOR AND DELIVERY

Amazingly, we still don't understand what triggers labor. Although the exact mechanisms continue to elude us, we do know something about the series of events that culminate in uterine contractions so powerful, they can propel a baby out into the world.

Roger Smith, director of the Mothers and Babies Research Centre at the University of Newcastle in Australia, likens the behavior of uterine muscle cells near birth to that of soccer fans. At any one time, many people may be clapping, but all we hear is random noise because there's no coordinated effort. Then a few fans in one section of the stadium begin to clap and sing together, chanting for their team. That effort may fade, only to be taken up by another group distant from the one who started the cheer. This process goes on for several minutes until the whole crowd is clapping and chanting in unison, producing an impressive effect.

A similar sequence of events happens in labor. Initially, uterine muscle cells aren't connected to one another. They may contract in isolation, but on their own they are powerless. As birth nears, connections form between groups of cells, and these islands begin to contract together. A group in one part of the uterus may be active for a time before it goes quiet and another patch of interconnected cells takes over. Soon many groups of cells are contracting. Eventually they coalesce, synchronizing their activity into one powerful unit, flexing its might in the process we call labor.

Preparation for delivery also includes gradual cervical changes. During most of pregnancy, the cervix provides mechanical support that helps hold the baby inside the uterus. It contains a gelatinous stopper of sorts, called the cervical mucus plug, which keeps microbes from reaching the baby. As birth approaches, the plug dissolves and the cervix undergoes a process called ripening. Its composition changes, becoming more pliant. Effacement refers to shortening and thinning of the cervix, and dilation denotes widening of its narrow opening into the vagina.

AM I REALLY IN LABOR?

Given the unorganized nature of those first uterine contractions and the gradual changes in the cervix leading to labor, it's no wonder that many women find it difficult to pinpoint the exact moment when the process starts. Only in retrospect is it clear what was going on. Here are some of the signs indicating that labor is near:

• There may be an increase in secretions from your vagina due to loss of the cervical mucus plug.

• Your uterus may tighten and relax (called Braxton-Hicks contractions), at first randomly and then on a more regular basis. This often feels like your abdominal muscles are suddenly squeezing your pregnant belly.

• You may have lower back pain.

• The amniotic sac ruptures (your water breaks), releasing fluid from the uterus into your vagina and producing a dribble or a flood.

If your water breaks, that's a sure sign you should call your obstetric care provider; you will likely be told to go to the hospital for evaluation. Based on the results, a plan will be made: admission or a return home to wait for stronger contractions. If labor does not begin spontaneously, you should ask how long to wait before returning to the hospital for induction.

Normal Labor and Vaginal Delivery

Clinicians divide labor into three stages. The **first stage** begins with labor contractions and ends with full cervical dilation. The start of labor contractions is marked from the time they come every three to five minutes over the course of an hour. The first stage ends when the cervix is fully dilated, creating an opening of about ten centimeters. You won't experience any momentous physical signs, such as changes in the pain level, that make it obvious that the cervix is fully dilated. A physical exam is the only way to know for sure.

The first stage of labor is subdivided into a **latent phase,** in which the cervix slowly dilates, and an **active phase,** when the pace quickens. If you've already had a baby, the transition between the two phases usually occurs when your cervical opening reaches about five centimeters. For first-time mothers, this may happen at a later stage, six centimeters or more. But for some women, whether or not they've given birth before, the transition may be muddled.

During the latent phase, it can take up to six hours for the cervix to dilate from four to five centimeters and about half that time to open just one centimeter more. Once the cervix is dilated to at least six centimeters, the active phase begins. The rate of change quickens to about one to two centimeters per hour for a woman who goes on to have a normal vaginal delivery. A persistently slower rate may indicate a problem.

Data suggest that the first stage of labor lasts longer today than in the past. One contributor is the increase in the use of epidural anesthesia, which can slow things down a bit. But this is not the whole story. Obesity can play a significant role, lengthening the active phase by over two hours. In addition, obstetricians aim for fewer interventions, letting nature take its course whenever possible.

The **second stage** of labor begins with full cervical dilation and ends with birth. During this stage the baby descends or rotates into an optimal position. This usually takes a little over half an hour if you have not had a baby before and about fifteen minutes if you have. Once again epidural anesthesia can lengthen the time. The second stage of labor ends with pushing and delivery of the baby.

The **third stage** of labor is the interval between birth and the delivery of the placenta. Once the baby is born, the uterus rapidly shrinks in volume. As its original attachment site becomes smaller and smaller, the placenta loses its foothold and is torn from the uterus, from which it is expelled. The third stage of labor takes ten to twenty minutes, with most placentas delivered within thirty minutes of the baby's appearance.

The usual procedure is for the obstetric care provider to deliver the placenta by gently pulling on the umbilical cord, placing one hand on the mother's abdomen, and asking her to push. If you haven't delivered the placenta within thirty minutes or if there is a lot of bleeding, other methods will come into play. Sometimes having you push in a squatting position will work, or your obstetric care provider may perform a manual extraction. Once delivered, the placenta is carefully examined to make sure that it emerged intact. Pieces that remain embedded in the uterine wall and its blood vessels can cause severe blood loss.

Even women who have already had a baby are often surprised to learn about the various transitions that mark the different stages of labor. Since you can't examine your own cervix or determine the baby's position, it's impossible to know about your progress in real time. Instead you'll get the news secondhand, at irregular intervals, as your obstetric care provider performs a physical exam. As a result, the whole process may seem like a blur, climaxing in the birth of your baby and ending with the delivery of the placenta. Knowing the landmarks along the road can help you chart your progress in what can be a confusing landscape.

WHY LABOR MAY BE INDUCED

For approximately a fifth of pregnant women in the United States, labor has to be induced. Here are some common reasons:

* A problem may arise that makes it riskier for the baby to stay in the uterus than to be born.
* Labor fails to begin at term.
* The placenta is at the end of its life span, which is matched to the length of pregnancy. If your pregnancy continues beyond your due date, placental function may decline, putting your baby in danger of losing its lifeline to you.

Generally, doctors concur that labor shouldn't be induced before thirty-nine weeks, which allows the baby to reach full maturity.

Induction simulates the sequence of events that occurs in natural labor — cervical changes followed by orchestrated uterine contractions. The method used depends on the extent to which your cervix has ripened, the process of softening prior to effacement (or shortening and thinning), and the extent of dilation (or opening). If the ripening process has yet to begin or is in the initial stages, induction starts by readying the cervix for birth. The most common approach involves insertion of a balloon-like device to mechanically initiate the opening process.

Once cervical ripening has progressed to a point deemed favorable for a vaginal delivery, induction of labor is commenced. Pitocin, the brand name for a synthetic version of the naturally occurring hormone oxytocin, is the drug of choice. Oxytocin's ability to stimulate uterine contractions was discovered early in the twentieth century (*oxytocin* comes from Greek roots meaning "sudden delivery"). Usually, the administration of Pitocin is followed by artificial rupture of the membranes: a hooklike device is used to make a small opening in the amniotic sac, a process that can be uncomfortable. This combination approach has been shown to be more effective than the use of either method alone.

Although it may take more time to get there, once women who have undergone induction enter active labor, the progress toward birth is no different than that of spontaneous labor.

TAKE CHARGE: MAXIMIZE YOUR CHANCES OF A SUCCESSFUL LABOR (AND MINIMIZE INTERVENTIONS)

Given the highly variable duration of the three stages of labor, it is hard to predict how long it will take any baby to be born. Parents' comments after the fact often reflect this lack of control: "My son took his sweet time — he was in no rush to enter the world!" "Our daughter is always in a hurry — with her, we barely made it to the hospital."

Even if you can't absolutely control what happens during labor and delivery, there are some choices you can make that might help the process along (no guarantees, though!).

These strategies apply *only if* you don't have pregnancy complications or medical problems.

Keep Your Support Person Close By

High on the list of factors that maximize a woman's satisfaction with a birth is limiting medical interventions. Having a reliable (and assertive) support person at your side during labor can help achieve this goal. Being in a hospital or birth center where high-tech medical care is available does not mean you have to accept being treated like a "typical patient," with all the poking, prodding, and adhering to rules this may entail. This is where your birth plan comes into play — for instance, maybe you want to stand or walk around rather than labor in a hospital bed.

Consider Adding an Experienced Professional to Your Team

Many women report that the continuous support of a knowledge-able person is especially crucial to managing labor and delivery successfully. Enlisting an experienced doula or a midwife, along with your chief supporter (who is likely to be a relative novice) can help. A person who has been through this process with many other women can help in ways that busy doctors and nurses focused on the medical aspects of your care may not have time for. A doula can also offer much-needed assurance that what you are experiencing is normal. Studies show that women who receive such continuous support are more likely to have a normal vaginal delivery and reduced rates of C-sections.

Exercise Your Freedom of Choice

You should feel free to opt for anesthesia; no evidence suggests that it substantially alters the course of labor and delivery. And a "no" decision does not have to be set in stone. You can change your mind as labor progresses without feeling you have somehow failed by not being able to handle the pain.

How to achieve the right amount of fluid intake is another area in which you have a choice. Hydration is important because dehydration can slow labor down. But this does not necessarily mean an IV. Some clinicians want to insert an IV in case this route is needed later for another reason, such as delivery of a drug. But others are comfortable with your drinking clear liquids during labor. The key question is whether you want to take responsibility for staying hydrated or let an IV do it for you.

Food is another matter to consider. Most doctors recommend that you not eat solid food during labor. If what you ate comes back up later, it could go down the "wrong tube" into the lungs, a condition called aspiration, which can cause serious problems just when you don't need them. But, depending on the hospital, you might be allowed to have solid food during the early stages of labor.

Get into the Position That Works for *You*

If everything during labor is going "by the book," you should be allowed to choose a comfortable position and to move around at will, if you find it helpful.

Breathe Easy

As with laboring, there is no single "right" way to breathe during contractions. There are commonly accepted strategies that are said to help labor progress, such as holding your breath as the pain is rising and letting it out in short bursts at the peak. But there is no evidence that one method works better than another.

Check Up on Your Baby

Fetal heart rate monitoring is a critical tool for obtaining real-time information about the baby's status. Normally its heart beats faster with each contraction. Decelerations (or the heart rate slowing) indicate that the baby's coping mechanisms may be failing. The standard practice is to monitor the heart rate at least every fifteen minutes during the active phase of the first stage of labor and every five minutes during the second. Depending on your preference, this can be done intermittently or continuously. Some women find the constant "data feed" and being "wired up" distracting, while others find it reassuring. One factor to consider is that staff are often too busy to do intermittent monitoring at the recommended intervals. This makes continuous monitoring the safer choice. And with modern technology, you don't have to stay in bed if you opt for this method.

Do It Your Way

Pushing and delivery can also be done in any position. Scream and yell if it helps. Some clinicians advise pushing with a closed glottis, which is the opening in the back of your throat that we naturally obstruct when lifting a heavy object or having a bowel movement. But if pushing with an open glottis comes naturally for you, then that's fine too. Instinct should trump technique because evidence does not favor a particular method. There's one exception — if the baby is born too quickly, the perineum (the area between the vagina and the anus) can tear. If this risk arises, you will be asked to pant in little breaths, to avoid pushing. This lets the baby emerge more slowly. Making a surgical incision (an episiotomy) to prevent a tear is no longer recommended except when required to prevent a serious laceration.

WHEN LABOR IS TOO SLOW

Sometimes labor does not progress normally. Instead of building to the crescendo that culminates in birth, it slows down or stops altogether. Unfortunately, abnormal progression is not uncommon, though the reasons for it are poorly understood. About 20 percent of all women experience one of these problems, and the rates can be substantially higher for first-time mothers. A slowdown in the pace of labor does not automatically mean a cesarean section is indicated, however. Most obstetric care providers will allow you to continue down the path to a vaginal delivery if the baby shows no sign of distress and progress is being made.

During the active phase of the first stage of labor, when the cervix is dilated six centimeters or more, the muscles of the uterus can fail to work together and contract with enough force to complete cervical dilation. Remember the soccer crowd analogy? Imagine that the crowd is only weakly enthusiastic or doesn't cheer in unison. A diagnosis of protraction is made when progress during the active phase of labor slows to below one to two centimeters per hour. If the rate decreases even further, to one centimeter or less every two hours, then Pitocin is usually administered to stimulate more effective contractions.

A diagnosis of arrest is made in the active phase of labor when the membranes have ruptured and four hours pass with no cervical changes. If, during this entire time, contractions are judged to have been sufficiently strong to produce dilation, then a cesarean section (see below) is the next step. If contractions have been weaker, then labor may be allowed to continue for another two hours before the baby is delivered surgically.

If the second stage lasts too long, a diagnosis of prolonged labor is made. From here on, the exact timing around the decision to do a surgical delivery depends on safety considerations for both the mother and soon-to-be-born child. If the baby is in the right position, progress may be slowed if the head is too large to easily slip out of the birth canal. Currently, the Society for Maternal-Fetal Medicine recommends letting women who've given birth before push for up to two hours from the start of the second stage of labor and first-time mothers an hour more (i.e., three hours). In each case, another hour is added on for epidural anesthesia. Along the way, Pitocin may be administered to help move things along. Usually, after the times indicated, lack of progress results in a cesarean section.

Cesarean Sections

Some cesarean sections are planned ahead of time, before labor begins, for medical reasons. Perhaps the placenta is implanted toward the bottom rather than the top of the uterus (see chapter 2), or the baby is in a breech position or is too large to pass through the mother's birth canal.

In other cases, the need for a cesarean section arises during labor. One cause is a failure to progress normally through the first or second stage. In some cases, the umbilical cord ends up in the wrong position and is pinched off as the baby descends toward the vagina. Another reason is that the mismatch between a large baby and a too-small birth canal may not be evident until labor is well under way. Or the baby may respond poorly to the birth process. The tracings on the heart monitor may show a decreased rate rather than the expected increase at the peak of a contraction. Any of these problems can trigger an emergency, or "crash," cesarean section.

The type of anesthesia offered depends on the circumstances. For an elective cesarean section, a powerful numbing medication is commonly administered to the fluid that surrounds the spinal cord (spinal anesthesia) or you may have an epidural. When the need for a surgical birth arises during labor, anesthesia may be delivered through the tubing that was previously placed for an epidural. Both techniques deaden sensation in the area where an incision will be made. General anesthesia is limited to emergency cesarean deliveries when spinal or epidural anesthesia cannot be performed or has failed.

A transverse or side-to-side incision (called a Pfannenstiel or "bikini") is the most common type because the scar is less noticeable than that of a vertical incision. The opening is made just above the pubic bone. Also this kind of incision heals better and creates a stronger junction of tissues, meaning that there is less of a chance that you will develop a hernia in the future. (A hernia occurs when portions of the intestines or other organs protrude through a weakened abdominal wall.) Pfannenstiel incisions are also less painful. But sometimes obesity or the need to deliver the baby as quickly as possible mandates a vertical cut.

If you have a C-section, your hospital stay may be extended (for example, three days rather than the one or two for a vaginal delivery). For a few weeks after the surgery, you should refrain from lifting anything heavier than your baby. Your clinician will give you a time line for what you can expect, including pain management, follow-up care, and the best time to resume certain activities that may be unsafe for a while. Take full advantage of the help you will be offered at the hospital, and make sure you have assistance at home. You're recovering from major abdominal surgery — and you have a new baby! If all goes well, you can expect a full recovery from your incision in about six weeks.

If you eventually become pregnant again after having a C-section, you may be wondering about your chances for a normal delivery (called a VBAC — vaginal birth after cesarean). The short answer is it depends on the medical reasons for your C-section and whether your desire for a vaginal birth is supported by your caregiver and hospital. Many women who previously had C-sections but who experience normal labor the next time can successfully have a VBAC, especially if they are allowed a little extra time to progress through labor and if the baby is doing fine. If you want to try for a VBAC, make your wishes known when you begin your prenatal care.

YOUR BABY IS HERE

The child you've carried for about nine months is out of the womb and in the world. Finally you can see, touch, and hold your baby. Decades after having my own children, I am still at a loss for words that come even remotely close to describing this experience. The momentous occasion is over, but there are still a few things you should know about the birthing process and some decisions to make about your newborn.

The Apgar Score

Generations ago, when powerful drugs such as morphine were routinely administered in copious amounts to women during the birth process, newborns were also affected because these agents crossed the placenta. Doctors found it challenging to distinguish between babies who were successfully emerging from the fog of the drugs and newborns who were truly struggling outside the womb. That's why the Apgar scoring system was developed.

The numerical system for evaluating how well a baby transitions from the womb to the world is named for its inventor, Dr. Virginia Apgar. In the late 1930s and early '40s, she trained first in surgery and then in anesthesia, which was just gaining recognition as a medical specialty. As a member of the faculty at the Columbia University School of Medicine (where she eventually became the first female professor), her research interest was obstetric anesthesia, which provided pain relief for mothers but could leave newborn babies groggy, making it harder for them to breathe on their own. She saw a need for a systematic way of identifying the ones who would need resuscitation, and in the early 1950s she invented one, which is used to this day.

There are five parts to the Apgar test, and each is graded on a scale of 0 to 2, with 10 a perfect score. Typically evaluations are made and recorded at one and five minutes after birth.

* *Color* is assessed. Pink, a sign of good oxygen delivery to the baby's body, receives a 2. Blue, indicative of poor oxygenation, is a 0.
* A *heart rate* of greater than a hundred beats per minute earns the highest mark. Lower rates receive lower scores.
* *Reflex irritability, muscle tone, and respiration* (2 points each) are also evaluated. A squirming baby who has no trouble breathing independently, as demonstrated by healthy cries of protest concerning what has just happened, is ideal.

Now, decades after Apgar scores were first developed, mountains of data exist about them. A mark below 5 at five minutes after birth may indicate a heightened risk of brain injury and other complications, such as cerebral palsy, due to lack of oxygen. But this is far from a certainty, as many children with low Apgar scores show no ill effects. A mark above 7 is reassuring; the likelihood of brain damage is low.

When to Cut the Umbilical Cord

The timing of "cutting the cord" can affect the health of your baby. Waiting thirty to sixty seconds after delivery to clamp the umbilical cord allows it to transfuse some of your newborn's blood, which happened to be in the placenta, back into the baby's circulation.

Studies show that babies whose cord clamping was delayed have more of the protein that carries oxygen in the blood (hemoglobin), higher iron levels a few months later, and better development. But if your baby needs help at this time (for example, resuscitation), the benefits of cutting the cord immediately may outweigh those of receiving a little extra placental blood.

Take the Stem Cells to the Bank?

Some parents decide to take the precaution of storing whatever residual blood can be drained from the severed cord. The stem cells from this cord blood could be used in a bone marrow transplant, should your child be diagnosed with a disease such as leukemia in the future. Saving and storing cord blood is expensive. Typical processing fees are about $1,500, and there are ongoing storage costs of around $200 per year. Currently ACOG and the Lymphoma and Leukemia Society recommend against for-profit umbilical cord banking because of the very low probability that it will ever be used. But there is an exception — when a sibling of the baby has such a disease, which could be treated by transplanting these cells. Another consideration is that for such a transplant to be effective, a bigger child needs a larger dose of stem cells, and umbilical cord blood contains relatively few; later in your child's life, the banked cells would need to be combined with cells from another source.

The immediate health benefits of letting the cord blood return to the baby's circulation outweigh the theoretical future benefits of banking, which are very unlikely to materialize. On the other hand, contributing your baby's umbilical cord blood to a public bank is a generous act that could benefit others.

DON'T GET TOO ATTACHED

Usually, the umbilical cord, like the placenta to which it is connected, receives very little attention after birth. But sometimes it receives too much.

Certain mothers are opting for a "Lotus Birth," in which the umbilical cord is not severed but rather allowed to separate naturally, which takes about a week, give or take a few days. During this time the placenta and the cord remain attached to the baby.

Having a Lotus Birth requires a few supplies—a carrying bag to tote the placenta and cord everywhere that the baby goes, and mixtures of dried herbs to mask their smell as they slowly decay. These items are easily obtained online (from Etsy, for instance)—if you truly want to go this route.

My advice: cut that cord! A Lotus Birth has no health benefits for the baby. In fact, it heightens the risk for health problems. The decomposing placenta can foster the growth of microbes, which may infect your newborn and cause a serious illness that was entirely avoidable.

THE PLACENTA: AFTER BIRTH

During the third stage of labor, you'll have a second set of contractions and deliver the placenta. It's much faster than delivering your baby (and much easier, given how comparatively small it is). By the time the placenta is expelled, your baby will have been out of the womb for ten to twenty minutes.

Not that long ago, no one paid any attention to the placenta once it emerged from the womb. Placentas were usually discarded as medical waste or spirited away and stored en masse, prior to being turned over to pharmaceutical companies as a source of products such as hormones. Some companies still put placental extracts in cosmetics. It's easy to find placenta-containing shampoo on the hard-to-reach shelves of drug stores, supposedly beneficial for the hormones and growth-promoting factors it contains. (In reality it's a marketing ploy to sell an ordinary shampoo with no special powers.)

Seen through the lens of time, this loss of interest in the placenta seems to be only a temporary dip. As far back as recorded history goes, the placenta was accorded the respect due to this most enigmatic of organs. When the biological basis of birth, life, and death was entirely shrouded in mystery, the role of the placenta must have seemed truly wondrous.

As with many things we humans don't understand, the placenta has been subjected to thousands of years of storytelling in an attempt to explain its existence. Some cultures viewed the placenta as a mother, sibling, companion, or doppelganger. Ancient Egyptians believed in the duality of the soul. One part inhabited the baby and the other the placenta. Some monarchs were drawn with a standard depicting their placentas. Other cultures had a more abstract view of the placenta as a spiritual force or having magical powers.

A logical extension of the personification and even glorification of the placenta was its ritual burial, which, in many parts of the world, became highly evolved. Placentas were placed in specially designed receptacles and interred in elaborate ceremonies. In some cultures, the final resting place was the sea, and in others, under a rosebush — any spot that was deemed auspicious for the child's life.

Today, most cultures no longer accord personhood or supernatural powers to the placenta. But in some parts of the world, respect remains, reflected in the desire for some sort of ritual marking the end of its life.

In the United States we commemorate people or significant events with the planting of a tree. Some families, in a similar spirit, have adopted the practice of burying their baby's placenta under a sapling.

Other more recent trends regarding disposition of the placenta involve less in the way of symbolism. Placentophagy, or eating the placenta, falls into this category. The rationale for doing this is based on the premise that we are animals, and other animals eat their placenta. But by the time your baby is born, the placenta has reached the end of its life span. Moreover, after its vaginal birth, it has taken an unusual path to your plate. Other sources of food are safer and more palatable.

A small for-profit industry has sprung up around putting a baby's placenta into capsules, which for many people makes ingesting it easier. Manufacturers make claims about the nutritional value of the organ and other benefits, such as a putative ability to ward off postpartum depression. These supposed advantages are unsubstantiated. What has been proven is that, in some cases, the processing does not destroy the microbes the postpartum placenta harbors, which can make the mother sick. In turn, she can transmit the illness to her baby. For this reason, the CDC issued a warning against taking placenta capsules.

Here's an interesting way to commemorate the placenta. My colleague and I made placenta prints, which we turned into surprisingly beautiful T-shirts for our research group. Online you can find many easy-to-follow instructional videos that tell you about the steps in the process and the supplies you will need to bring to the hospital, including the type of paper and ink. If you (or, more realistically, your support person) want to turn the placenta into a work of art, be sure to include printmaking in your birth plan. Also, alert your obstetric care providers so that the placenta is not inadvertently incinerated as medical waste.

As a researcher, I am very grateful to all the women and their families who have donated placentas to our group. Without their generosity, our work would not be possible. Studying the placenta is helping us unlock the mysteries of pregnancy, and we hope that doing so will make it a safer process for women and their babies.

My recommendation? If this option is possible, donate your baby's placenta to science. You'll be making a solid investment in the future.

YOUR PREGNANCY, YOUR BABY — YOU'RE IN CHARGE

It seems fitting that this book — which began with the role of placental cells in embryo implantation, the critical first step in establishing pregnancy — concludes with delivery of the placenta, the third stage of labor and the end of pregnancy.

It is my sincere hope that by understanding how pregnancy begins and ends and what happens in between, you will be able to use this knowledge to optimize your chances of having a healthy baby. There are many things that you can do to maximize your chances of success. Yes, pregnancy is a natural process, but this does not mean that you should be a passive bystander.

I strongly believe that, had I known some of the information presented in this book, I could have avoided the pitfalls I encountered during my first pregnancy, including preterm labor. After our first baby was born, I researched its causes and learned that my lifestyle, which included standing on my feet in a laboratory for long stretches of time, could be a contributing factor. Pregnant with my second child, I made a point of resting for a time every day, which I think helped me avoid another bout of preterm labor. When it came to methods of pain control during labor, I knew that I wanted an epidural, which transformed my second delivery.

In addition to learning from my own experiences, I have now spent a lifetime studying pregnancy. A lot has happened between the time when I had my children and now. We still don't know as much as we should about human pregnancy, but we are learning. The field has advanced, and it is important to take advantage of what we know today to make your pregnancy journey as safe as possible for you and your baby.

RESOURCES

CHAPTER 1

Pregnancy registries: https://www.fda.gov/ScienceResearch/SpecialTopics/
WomensHealthResearch/ucm251314.htm

Zika information: https://www.cdc.gov/zika/index.html

CHAPTER 4

CDC National Biomonitoring Program, list of environmental chemicals:
https://www.cdc.gov/exposurereport/index.html

Environmental Working Group shopping guides: http://www.ewg.org/
foodnews

Flame retardant chemicals buying guides: https://www.ceh.org

General information on environmental chemical exposures: http://
greensciencepolicy.org

Herbal remedies and alternative medicine safety: https://mothertobaby.org/?s
=herbal+remedies

Integrated pest-management strategies (gardening without pesticides): http://
ipm.ucanr.edu/index.html

Nontoxic personal-care products: https://www.ewg.org/skindeep

Pediatric Environmental Health Specialty Units: https://www.pehsu.net

Pesticide Action Network app: http://www.whatsonmyfood.org

Program on Reproductive Health and the Environment (PRHE): https://prhe
.ucsf.edu

Your rights in the workplace: https://prhe.ucsf.edu/work-matters-know-your
-rights

CHAPTER 5

Diabetes and women, general information: https://www.acog.org/
patient-resources/faqs/womens-health/diabetes-and-women

Gestational diabetes, management and treatment: https://www.niddk.nih
.gov/health-information/diabetes/overview/what-is-diabetes/gestational/
management-treatment

Preeclampsia Foundation, resources and information: https://www
.preeclampsia.org

Preterm birth data by state: https://www.marchofdimes.org/mission/
prematurity-reportcard.aspx

CHAPTER 7

Printable birth plan: https://www.marchofdimes.org/materials/birth-plan.pdf

ACKNOWLEDGMENTS

This book had a long gestation. The idea was the brainchild of the literary agent, Katherine "Kitty" Cowles, who read a feature story on the placenta in the science section of the *New York Times*. Our research group at the University of California San Francisco was prominently featured. Kitty contacted me, saying that she wanted to discuss the possibility of turning the piece into something bigger — a book. Would I be interested?

Her idea deeply resonated with me. Throughout my career, I have been acutely aware of how hard it is for most women to obtain accurate information about the biological basis of pregnancy. I know only too well what this experience feels like from my own experience with preterm labor. Between diagnosis of the condition and delivery of the baby, while I was on bed rest, I had many weeks to think about how much I wished there was a book that distilled the essential facts about my problem. By the time Kitty approached me, I had been mulling over this void for many years.

I am also grateful to Rux Martin, former editorial director of Rux Martin Books at Houghton Mifflin Harcourt (HMH), for bringing this project to life, and to Deb Brody, now vice president and publisher of adult trade at HMH, for taking over when Rux retired. Their work made the information more accessible and helped highlight the practical actions that pregnant women can take.

Nikola Kolundzic contributed the artwork to this book. His attention to detail, ability to grasp the underlying science, and creative talents are evident in his skillful renderings.

Several of my colleagues played important roles. Dr. Sig-Linda Jacobson, an obstetrician and maternal-fetal medicine subspecialist, has been my friend for decades. We first connected by sneaking out of boring sessions at international conferences to go exploring. I asked her to read the last chapter, on birth, which she did, then surprised me by offering to review more! I am grateful for her valuable input.

I want to thank Dr. Chris Redman, who was willing to review chapter 5, on pregnancy complications. Before retirement he was a clinical professor of medicine and a consultant to the Nuffield Department of Obstetrics and Gynecology (Oxford University). I first got to know Chris from his published papers, which are distinguished by scholarship and clarity. His wry humor and sense of mischief make him a favorite for confabs about pregnancy research or any other topics.

My colleagues at the University of California San Francisco also generously contributed to this book. Dr. Tracey Woodruff, professor and the director of the Program on Reproductive Health and the Environment, helped me with chapter 4. Tracey has dedicated her career to identifying, measuring, and preventing exposures to environmental contaminants that affect human reproduction and development.

Dr. Mike German, a professor in the Department of Medicine, treats patients with diabetes and researches the development of insulin-producing pancreatic beta cells. Mike helped me understand the many facets of gestational diabetes.

Dr. Mike McMaster is a professor and scientist who has been with my research group for nearly thirty years. During that time he has helped me edit numerous papers and grant applications. I am particularly grateful for his input on chapters 1 and 2, on the early stages of human development, his area of expertise.

Becky Cabaza did important editorial work on this book, helping me synthesize the chapters and explain the science clearly. I valued our regular conversations both professionally and personally.

The biggest thank-you goes to my family, the foundation of my life. My love for my husband grew out of a friendship based on a mutual fascination, or some might say obsession, with science. As reproductive biologists, we never doubted that we wanted to experience firsthand what we had studied. Before we had children, we had many humorous discussions about how they might turn out. What if the babies were a curious amalgam of our least desirable features? I am happy to say that just the opposite happened. Somehow, when we rolled the genetic dice, our two daughters inherited our best qualities. And to our great pleasure, they also love science.

REFERENCES

INTRODUCTION

Niederberger, C., Pellicer, A., Cohen, J., et al. 2018. "Forty years of IVF." *Fertil Steril* 110 (2): 185–324. doi: 10.1016/j.fertnstert.2018.06.005.

CHAPTER 1

American College of Obstetricians and Gynecologists. 2006. "Practice bulletin no. 75: Management of alloimmunization during pregnancy." *Obstet Gynecol* 108 (2): 457–64. doi: 10.1097/00006250-200608000-00044.

——2018. "Committee on Practice Bulletins—Obstetrics. Committee opinion no. 741: Maternal immunization." *Obstet Gynecol* 131(6):e214–e217. doi: 10.1097.AOG.0000000000002662.

——2017. "Committee opinion no. 721: Summary: Smoking cessation during pregnancy — Practice guidelines." *Obstet Gynecol* 130 (4): 929–30. doi: 10.1097/AOG.0000000000002348.

——2017. "Committee on Genetics. Committee opinion no. 690: Carrier screening in the age of genomic medicine: Practice guideline." *Obstet Gynecol* 129 (3): e35–e40. doi: 10.1097/AOG.0000000000001951.

Berard, A., Zhao, J. P., and Sheehy, O. 2016. "Success of smoking cessation interventions during pregnancy." *Am J Obstet Gynecol* 215 (5): 611e1–611e8.

Blount, B. C., Karwowski, M. P., Shields, P. G., et al. 2020. "Vitamin E acetate in bronchoalveolar-lavage fluid associated with EVALI." *N Engl J Med* 382 (8): 697–705.

Boston Children's Hospital. 2018. "TORCH." http://www.childrenshospital .org/conditions-and-treatments/conditions/t/torch.

Cooper, A. R., and Moley, K. H. 2008. "Maternal tobacco use and its preimplantation effects on fertility: More reasons to stop smoking." *Semin Reprod Med* 26 (2): 204–12.

Corsi, D. J., Walsh, L., Weiss, D., et al. 2019. "Association between self-reported prenatal cannabis use and maternal, perinatal, and neonatal outcomes." *JAMA* 322 (2): 145–52.

De Luca, M., Aiuti, A., Cossu, G., et al. 2019. "Advances in stem cell research and therapeutic development." *Nat Cell Biol* 21 (7): 801–11.

Ding, H., Black, C. L., Ball, S., et al. 2017. "Influenza vaccination coverage among pregnant women — United States, 2016–17 influenza season." *MMWR Morb Mortal Wkly Rep* 66 (38): 1016–22.

Dodge, L. E., Missmer, S. A., Thornton, K. L., et al. 2017. "Women's alcohol consumption and cumulative incidence of live birth following in vitro fertilization." *J Assist Reprod Genet* 34 (7): 877–83.

Dontigny, L., Arsenault, M. Y., and Martel, M. J. 2018. "No. 203: Rubella in pregnancy." *J Obstet Gynaecol Can* 40 (8): e615–e621.

Dryburgh, L. M., Bolan, N. S., Grof, C. P. L., et al. 2018. "Cannabis contaminants: Sources, distribution, human toxicity, and pharmacologic effects." *Br J Clin Pharmacol* 84 (11): 2468–76.

Edwards, K. M. 2017. "Ensuring vaccine safety in pregnant women." *N Engl J Med* 376 (13): 1280–82.

Faucette, A. N., Unger, B. L., Gonik, B., et al. 2015. "Maternal vaccination: Moving the science forward." *Hum Reprod Update* 21 (1): 119–35.

Feodor Nilsson, S., Andersen, P. K., Strandberg-Larsen, K., et al. 2014. "Risk factors for miscarriage from a prevention perspective: A nationwide follow-up study." *BJOG* 121 (11): 1375–84.

Grant, K. S., Petroff, R., Isoherranen, N., et al. 2018. "Cannabis use during pregnancy: Pharmacokinetics and effects on child development." *Pharmacol Ther* 182: 133–51.

Hart, R. J. 2016. "Physiological aspects of female fertility: Role of the environment, modern lifestyle, and genetics." *Physiol Rev* 96 (3): 873–909.

Hoyme, H. E., Kalberg, W. O., Elliott, A. J., et al. 2016. "Updated clinical guidelines for diagnosing fetal alcohol spectrum disorders." *Pediatrics* 138 (2): e20154256.

Ilic, D., and Ogilvie, C. 2017. "Concise review: Human embryonic stem cells — What have we done? What are we doing? Where are we going?" *Stem Cells* 35 (1): 17–25.

Moise, K. J., Jr., and Argoti, P. S. 2012. "Management and prevention of red cell alloimmunization in pregnancy: A systematic review." *Obstet Gynecol* 120 (5): 1132–39.

Moniz, M. H., and Beigi, R. H. 2014. "Maternal immunization: Clinical experiences, challenges, and opportunities in vaccine acceptance." *Hum Vaccin Immunother* 10 (9): 2562–70.

Moro, P., Baumblatt, J., Lewis, P., et al. 2017. "Surveillance of adverse events after seasonal influenza vaccination in pregnant women and their infants in the vaccine adverse event reporting system, July 2010–May 2016." *Drug Saf* 40 (2): 145–52.

Musso, D., Ko, A. I., and Baud, D. 2019. "Zika virus infection — After the pandemic." *N Engl J Med* 381 (15): 1444–57.

Myers, K. L. 2016. "Predictors of maternal vaccination in the United States: An integrative review of the literature." *Vaccine* 34 (34): 3942–49.

Neu, N., Duchon, J., and Zachariah, P. 2015. "TORCH infections." *Clin Perinatol* 42 (1): 77–103, viii.

Patorno, E., Huybrechts, K. F., Bateman, B. T., et al. 2017. "Lithium use in pregnancy and the risk of cardiac malformations." *N Engl J Med* 376 (23): 2245–54.

Pineles, B. L., Hsu, S., Park, E., et al. 2016. "Systematic review and meta-analyses of perinatal death and maternal exposure to tobacco smoke during pregnancy." *Am J Epidemiol* 184 (2): 87–97.

Pineles, B. L., Park, E., and Samet, J. M. 2014. "Systematic review and meta-analysis of miscarriage and maternal exposure to tobacco smoke during pregnancy." *Am J Epidemiol* 179 (7): 807–23.

Psychoyos, D., and Vinod, K. Y. 2013. "Marijuana, spice 'herbal high,' and early neural development: Implications for rescheduling and legalization." *Drug Test Anal* 5 (1): 27–45.

Richardson, K. A., Hester, A. K., and McLemore, G. L. 2016. "Prenatal cannabis exposure — The "first hit" to the endocannabinoid system." *Neurotoxicol Teratol* 58: 5–14.

Riley, E. R., Cahill, A. G., Beigi, R., et al. 2017. "Improving safe and effective use of drugs in pregnancy and lactation: Workshop summary." *Am J Perinatol* 34 (8): 826–32.

Sadler, T. W., and Langman, J. 2019. *Langman's medical embryology.* 14th ed. Philadelphia: Wolters Kluwer Health/Lippincott Williams & Wilkins.

Schwartz, D. A., and Dhaliwal, A. 2020. "Infections in pregnancy with COVID-19 and other respiratory RNA virus diseases are rarely, if ever, transmitted to the fetus: Experiences with coronaviruses, parainfluenza, metapneumovirus respiratory syncytial virus, and influenza." *Arch Pathol Lab Med* 144 (8).

Sharma, R., Harlev, A., Agarwal, A., et al. 2016. "Cigarette smoking and semen quality: A new meta-analysis examining the effect of the 2010 World Health Organization laboratory methods for the examination of human semen." *Eur Urol* 70 (4): 635–45.

Shavell, V. I., Moniz, M. H., Gonik, B., et al. 2012. "Influenza immunization in pregnancy: Overcoming patient and health care provider barriers." *Am J Obstet Gynecol* 207 (3 Suppl): S67–74.

Siu, A. L., and US Preventive Services Task Force. 2015. "Behavioral and pharmacotherapy interventions for tobacco smoking cessation in adults, including pregnant women: US Preventive Services Task Force Recommendation Statement." *Ann Intern Med* 163 (8): 622–34.

Soneji, S., and Beltran-Sanchez, H. 2019. "Association of maternal cigarette smoking and smoking cessation with preterm birth." *JAMA Netw Open* 2 (4): e192514.

Staud, F., Cerveny, L., and Ceckova, M. 2012. "Pharmacotherapy in pregnancy: Effect of ABC and SLC transporters on drug transport across the placenta and fetal drug exposure." *J Drug Target* 20 (9): 736–63.

Talbot, P., and Lin, S. 2011. "The effect of cigarette smoke on fertilization and pre-implantation development: Assessment using animal models, clinical data, and stem cells." *Biol Res* 44 (2): 189–94.

Thadani, P. V., Strauss, J. F. 3rd, Dey, S. K., et al. 2004. "National Institute on Drug Abuse Conference report on placental proteins, drug transport, and fetal development." *Am J Obstet Gynecol* 191 (6): 1858–62.

Theunissen, T. W., and Jaenisch, R. 2017. "Mechanisms of gene regulation in human embryos and pluripotent stem cells." *Development* 144 (24): 4496–509.

Zdravkovic, T., Genbacev, O., Mcmaster, M. T., et al. 2005. "The adverse effects of maternal smoking on the human placenta: A review." *Placenta* 26 (Suppl A): S81–S86.

Zenzes, M. T. 2000. "Smoking and reproduction: Gene damage to human gametes and embryos." *Hum Reprod Update* 6 (2): 122–31.

CHAPTER 2

Abramowicz, J. S., and Sheiner, E. 2008. "Ultrasound of the placenta: A systematic approach. Part I: Imaging." *Placenta* 29 (3): 225–40.

Abramowitz, A., Miller, E. S., and Wisner, K. L. 2017. "Treatment options for hyperemesis gravidarum." *Arch Womens Ment Health* 20 (3): 363–72.

American College of Obstetricians and Gynecologists. 2010. "Committee opinion no. 462: Moderate caffeine consumption during pregnancy." *Obstet Gynecol* 116 (2, Pt 1): 467–78. doi: 10.1097/AOG.0b013e3181eeb2a1.

———. 2013. "Committee opinion no. 548: Weight gain during pregnancy." *Obstet Gynecol* 121 (1): 210–12. doi: 10.1097/01.aog.0000425668.87506.4c.

———. 2015. "Morning sickness: Nausea and vomiting of pregnancy." In *Frequently asked questions — Pregnancy*, FAQ126. https://www.acog.org/Patients/FAQs/Morning-Sickness-Nausea-and-Vomiting-of-Pregnancy.

———. 2018. "Nutrition during pregnancy." *Frequently Asked Questions — Pregnancy*, FAQ001, 2–3, 313–27. https://www.acog.org/patient-resources/faqs/pregnancy/nutrition-during-pregnancy.

Benirschke, K., Burton, G. J., and Baergen, R. N. 2012. *Pathology of the human placenta*. Berlin and Heidelberg: Springer-Verlag.

Buck Louis, G. M., Grewal, J., Albert, P. S., et al. 2015. "Racial/ethnic standards for fetal growth: The NICHD fetal growth studies." *Am J Obstet Gynecol* 213 (4): 449 e1–449 e41.

Burton, G. J., Fowden, A. L., and Thornburg, K. L. 2016. "Placental origins of chronic disease." *Physiol Rev* 96 (4): 1 509–65.

Burton, G. J., and Jauniaux, E. 2015. "What is the placenta?" *Am J Obstet Gynecol* 213 (4 Suppl): S6 e1, S6–S8.

Burton, G. J., Jauniaux, E., and Charnock-Jones, D. S. 2010. "The influence of the intrauterine environment on human placental development." *Int J Dev Biol* 54 (2–3): 303–12.

Carberry, A. E., Gordon, A., Bond, D. M., et al. 2011. "Customised versus population-based growth charts as a screening tool for detecting small for gestational age infants in low-risk pregnant women" (review). *The Cochrane Collaboration.* https://www.cochrane.org/CD008549/PREG_customised -versus-population-based-growth-charts-as-a-screening-tool-for-detecting -small-for-gestational-age-infants-in-low-risk-pregnant-women.

Centers for Disease Control and Prevention. 2016. "Guidelines for vaccinating pregnant women." https://www.cdc.gov/vaccines/pregnancy/hcp-toolkit/ guidelines.html.

Cetin, I., Parisi, F., Berti, C., et al. 2012. "Placental fatty acid transport in maternal obesity." *J Dev Orig Health Dis* 3 (6): 409–14.

Clapp, C., Thebault, S., Jeziorski, M. C., et al. 2009. "Peptide hormone regulation of angiogenesis." *Physiol Rev* 89 (4): 1177–215.

Confavreux, C., Hutchinson, M., Hours, M. M., et al. 1998. "Rate of pregnancy-related relapse in multiple sclerosis: Pregnancy in multiple sclerosis group." *N Engl J Med* 339 (5): 285–91.

Crochet, J. R., Bastian, L. A., and Chireau, M. V. 2013. "Does this woman have an ectopic pregnancy? The rational clinical examination systematic review." *JAMA* 309 (16): 1722–29.

Dean, C. R., Shemar, M., Ostrowski, G. U., et al. 2018. "Management of severe pregnancy sickness and hyperemesis gravidarum." *BMJ* 363: k5000.

Donnez, J., and Dolmans, M. M. 2016. "Uterine fibroid management: From the present to the future." *Hum Reprod Update* 22 (6): 665–86.

Erlebacher, A. 2013. "Immunology of the maternal-fetal interface." *Annu Rev Immunol* 31: 387–411.

Erlebacher, A., and Fisher, S. J. 2017. "Baby's first organ." *Sci Am* 317 (4): 46–53.

Fejzo, M. S., Sazonova, O. V., Sathirapongsasuti, J. F., et al. 2018. "Placenta and appetite genes GDF15 and IGFBP7 are associated with hyperemesis gravidarum." *Nat Commun* 9 (1): 1178.

Fisher, S. J., Leitch, M. S., Kantor, M. S., et al. 1985. "Degradation of extracellular matrix by the trophoblastic cells of first-trimester human placentas." *Journal of Cellular Biochemistry* 27: 31–41.

Garner, C. D. 2017. "Nutrition in pregnancy." In C. J. Lockwood and D. Seres (Eds.), *UpToDate.* Waltham, MA: Wolters Kluwer Health Division of Wolters Kluwer. https://www.uptodate.com/contents/nutrition-in -pregnancy?search=Nutrition%20in%20pregnancy&source=search_result &selectedTitle=1~150&usage_type=default&display_rank=1.

Gellersen, B., and Brosens, J. J. 2014. "Cyclic decidualization of the human endometrium in reproductive health and failure." *Endocr Rev* 35 (6): 851–905.

Genbacev, O., Bass, K. E., Joslin, R. J., et al. 1995. "Maternal smoking inhibits early human cytotrophoblast differentiation." *Reprod Toxicol* 9 (3): 245–55.

Genbacev, O. D., Prakobphol, A., Foulk, R. A., et al. 2003. "Trophoblast L-selectin-mediated adhesion at the maternal-fetal interface." *Science* 299 (5605): 405–8.

Giudice, L. C. 2016. "Challenging dogma: The endometrium has a microbiome with functional consequences!" *Am J Obstet Gynecol* 215 (6): 682–83.

Handwerger, S., and Freemark, M. 2000. "The roles of placental growth hormone and placental lactogen in the regulation of human fetal growth and development." *J Pediatr Endocrinol Metab* 13 (4): 343–56.

Hay, W. W., Jr. 1994. "Placental transport of nutrients to the fetus." *Horm Res* 42 (4–5): 215–22.

Hromatka, B. S., Tung, J. Y., Kiefer, A. K., et al. 2015. "Genetic variants associated with motion sickness point to roles for inner ear development, neurological processes, and glucose homeostasis." *Hum Mol Genet* 24 (9): 2700–708.

Jauniaux, E., Alfirevic, Z., Bhide, A. G., et al. 2019. "Placenta praevia and placenta accreta: Diagnosis and management — Green-top Guideline No. 27a." *BJOG* 126 (1): e1–e48.

Johnston, Christine, et al. 2016. "Status of Vaccine Research and Development of Vaccines for Herpes Simplex Virus." *Vaccine* 34 (26): 2948–52. doi:10.1016/j.vaccine.2015.12.076.

Joshi, A. A., Vaidya, S. S., St.-Pierre, M. V., et al. 2016. "Placental ABC transporters: Biological impact and pharmaceutical significance." *Pharm Res* 33 (12): 2847–78.

Joshi, B., Aggarwal, N., and Singh, A. 2014. "Obstetrical catastrophe averted: Successful outcome of an abdominal pregnancy." *Am J Emerg Med* 32 (10): 1299, e3–e4.

Lee, N. M., and Saha, S. 2011. "Nausea and vomiting of pregnancy." *Gastroenterol Clin North Am* 40 (2): 309–34, vii.

Levine, S. Z., Kodesh, A., Viktorin, A., et al. 2018. "Association of maternal use of folic acid and multivitamin supplements in the periods before and during pregnancy with the risk of autism spectrum disorder in offspring." *JAMA Psychiatry* 75 (2): 176–84.

Lewis, R. M., and Desoye, G. 2017. "Placental lipid and fatty acid transfer in maternal overnutrition." *Ann Nutr Metab* 70 (3): 228–31.

Linzer, D. I., and Fisher, S. J. 1999. "The placenta and the prolactin family of hormones: Regulation of the physiology of pregnancy." *Mol Endocrinol* 13 (6): 837–40.

Loke, Y. W. 2013. *Life's vital link: The astonishing role of the placenta.* 1st ed. Oxford, UK: Oxford University Press.

Looker, A. C., Johnson, C. L., Lacher, D. A., et al. 2011. "Vitamin D status: United States, 2001–2006." *NCHS Data Brief* (59): 1–8.

Maltepe, E., and Fisher, S. J. 2015. "Placenta: The forgotten organ." *Annu Rev Cell Dev Biol* 31 (1): 523–52.

Matthews, A., Haas, D. M., O'Mathuna, D. P., et al. 2015. "Interventions for nausea and vomiting in early pregnancy." *Cochrane Database Syst Rev* CD007575. doi: 10.1002/14651858.CD007575.pub4.

Michelsen, T. M., Holme, A. M., and Henriksen, T. 2017. "Transplacental nutrient transfer in the human in vivo determined by 4 vessel sampling." *Placenta* 59 (Suppl 1): S26–S31.

Mirzakhani, H., Litonjua, A. A., McElrath, T. F., et al. 2016. "Early pregnancy vitamin D status and risk of preeclampsia." *J Clin Invest* 126 (12): 4702–15.

Mittal, A., Pachter, L., Nelson, J. L., et al. 2015. "Pregnancy-induced changes in systemic gene expression among healthy women and women with rheumatoid arthritis." *PLoS One* 10 (12): e0145204.

Mohamed, S. A., Al-Hendy, A., Schulkin, J., et al. 2016. "Opinions and practice of US-based obstetrician-gynecologists regarding vitamin D screening and supplementation of pregnant women." *J Pregnancy* 2016: 1454707.

Mold, J. E., and McCune, J. M. 2012. "Immunological tolerance during fetal development: From mouse to man." *Adv Immunol* 115: 73–111.

Moreno, I., Codoner, F. M., Vilella, F., et al. 2016. "Evidence that the endometrial microbiota has an effect on implantation success or failure." *Am J Obstet Gynecol* 215 (6): 684–703.

Mumford, S. L., Garbose, R. A., Kim, K., et al. 2018. "Association of preconception serum 25-hydroxyvitamin D concentrations with live birth and pregnancy loss: A prospective cohort study." *Lancet Diabetes Endocrinol* 6 (9): 725–32.

Murthi, P., Yong, H. E., Ngyuen, T. P., et al. 2016. "Role of the placental vitamin D receptor in modulating feto-placental growth in fetal growth restriction and preeclampsia-affected pregnancies." *Front Physiol* 7: 43.

Myers, K. M., and Elad, D. 2017. Biomechanics of the human uterus. *Wiley Interdiscip Rev Syst Biol Med* 9 (5): e1388. doi: 10.1002/wsbm.1388.

Myllynen, P., Immonen, E., Kummu, M., et al. 2009. "Developmental expression of drug metabolizing enzymes and transporter proteins in human placenta and fetal tissues." *Expert Opin Drug Metab Toxicol* 5 (12): 1483–99.

Nelson, D. M. 2015. "How the placenta affects your life, from womb to tomb." *Am J Obstet Gynecol* 213 (4 Suppl): S12–3.

Palacios, C., Kostiuk, L. K., and Pena-Rosas, J. P. 2019. "Vitamin D supplementation for women during pregnancy." *Cochrane Database Syst Rev* 7: CD008873. doi: 10.1002/14651858.CD008873.pub4.

Piccinni, M. P., Lombardelli, L., Logiodice, F., et al. 2016. "How pregnancy can affect autoimmune diseases progression?" *Clin Mol Allergy* 14: 11.

Robinson, J. F., Kapidzic, M., Hamilton, E. G., et al. 2019. "Genomic profiling of BDE-47 effects on human placental cytotrophoblasts." *Toxicol Sci* 167 (1): 211–26.

Rosso, M., Sijanovic, S., Topolovec, Z., et al. 2014. "Secondary abdominal appendicular pregnancy: Case report." *Srp Arh Celok Lek* 142 (7–8): 484–87.

Sanu, O., and Lamont, R. F. 2011. "Hyperemesis gravidarum: Pathogenesis and the use of antiemetic agents." *Expert Opin Pharmacother* 12 (5): 737–48.

Schatz, F., Guzeloglu-Kayisli, O., Arlier, S., et al. 2016. "The role of decidual cells in uterine hemostasis, menstruation, inflammation, adverse pregnancy outcomes, and abnormal uterine bleeding." *Hum Reprod Update* 22 (4): 497–515.

Tal, R., Taylor, H. S., Burney, R. O., et al. 2015. "Endocrinology of pregnancy." In K. R. Feingold and R. Rebar, (Eds.), *Endotext*. South Dartmouth, MA: MDText.com.

Thornburg, K. L., Jacobson, S. L., Giraud, G. D., et al. 2000. "Hemodynamic changes in pregnancy." *Semin Perinatol* 24 (1): 11–14.

Tsevat, D. G., Wiesenfeld, H. C., Parks, C., et al. 2017. "Sexually transmitted diseases and infertility." *Am J Obstet Gynecol* 216 (1): 1–9.

Vukusic, S., and Marignier, R. 2015. "Multiple sclerosis and pregnancy in the 'treatment era.'" *Nat Rev Neurol* 11 (5): 280–89.

Whitley, Richard J, and Bernard Roizman. 2001. "Herpes Simplex Virus

Infections." *Lancet* 357 (9267): 1513–18. doi:10.1016/s0140-6736(00) 04638-9.

Wiesenfeld, H. C. 2017. "Screening for *Chlamydia trachomatis* infections in women." *N Engl J Med* 376 (22): 2198.

Wilcox, A. J., Weinberg, C. R., O'Connor, J. F., et al. 1988. "Incidence of early loss of pregnancy." *N Engl J Med* 319 (4): 189–94.

Zondervan, K. T., Becker, C. M., and Missmer, S. A. 2020. "Endometriosis." *N Engl J Med* 382 (13): 1244–56.

CHAPTER 3

Alfirevic, Z., Navaratnam, K., and Mujezinovic, F. 2017. "Amniocentesis and chorionic villus sampling for prenatal diagnosis." *Cochrane Database Syst Rev* 9: CD003252. doi: 10.1002/14651858.CD003252.pub2.

Alldred, S. K., Takwoingi, Y., Guo, B., et al. 2017. "First and second trimester serum tests with and without first trimester ultrasound tests for Down's syndrome screening." *Cochrane Database Syst Rev* 3 (3): CD012599. doi: 10.1002/14651858.CD012599.

Antiel, R. M., and Flake, A. W. 2017. "Responsible surgical innovation and research in maternal-fetal surgery." *Semin Fetal Neonatal Med* 22 (6): 423–27.

Benacerraf, B. R. 2019. "Sonographic findings associated with fetal aneu- ploidy." In L. Wilkens-Haug and D. Levine (Eds.), *UpToDate*. Waltham, MA: Wolters Kluwer Health Division of Wolters Kluwer. https://www .uptodate.com/contents/sonographic-findings-associated-with-fetal -aneuploidy.

Benn, P. 2016. "Expanding non-invasive prenatal testing beyond chromosomes 21, 18, 13, X and Y." *Clin Genet* 90 (6): 477–85.

Bianchi, D. W., and Chiu, R. W. K. 2018. "Sequencing of circulating cell-free DNA during pregnancy." *N Engl J Med* 379 (5): 464–73.

Bianchi, D. W., Parker, R. L., Wentworth, J., et al. 2014. "DNA sequencing versus standard prenatal aneuploidy screening." *N Engl J Med* 370 (9): 799–808.

Biggio J. R., Kuller, J. A., and Blackwell, S. C. 2017. The role of ultrasound in women who undergo cell-free DNA screening. *Am J Obstet Gynecol* 216 (3): B2–B7. doi: 10.1016/j.ajog.2017.01.005.

Brady, P., Brison, N., Van Den Bogaert, K., et al. 2016. "Clinical implementation of NIPT — Technical and biological challenges." *Clin Genet* 89 (5): 523–30.

Canick, J. 2012. "Prenatal screening for trisomy 21: Recent advances and guidelines." *Clin Chem Lab Med* 50 (6): 1003–8.

Cargill, Y., and Morin, L. 2017. "No. 223–Content of a complete routine second trimester obstetrical ultrasound examination and report." *J Obstet Gynaecol Can* 39 (8): e144–e149.

Chard, R. L., and Norton, M. E. 2016. "Genetic counseling for patients considering screening and diagnosis for chromosomal abnormalities." *Clin Lab Med* 36 (2): 227–36.

Committee on Genetics and the Society for Maternal-Fetal Medicine. 2016. "Committee opinion no. 682 (6): Microarrays and next-generation sequencing technology — The use of advanced genetic diagnostic tools in obstetrics and gynecology." *Obstet Gynecol* 128 (6): e262–e268. doi: 10.1097/AOG.0000000000001817.

Committee on Obstetric Practice. 2017. "Committee opinion no. 723: Guidelines for diagnostic imaging during pregnancy and lactation." *Obstet Gynecol* 130 (4): e210–e216. doi: 10.1097/AOG.0000000000002355.

Committee on Practice Bulletins—Obstetrics and the American Institute of Ultrasound in Medicine. 2016. "Practice bulletin no. 175: Ultrasound in pregnancy." *Obstet Gynecol* 128 (6): e241–e256. doi: 10.1097/AOG.0000000000001815.

Committee on Practice Bulletins—Obstetrics, Committee on Genetics, and the Society for Maternal-Fetal Medicine. 2016. "Practice bulletin no. 163: Screening for fetal aneuploidy." *Obstet Gynecol* 127 (5): e123–e137. doi: 10.1097/AOG.0000000000001406.

Cook-Deegan, R., and Chandrasekharan, S. 2016. "*Sequenom v. Ariosa*: The death of a genetic testing patent." *N Engl J Med* 375 (25): 2418–19.

Cuckle, H., and Maymon, R. 2016. "Development of prenatal screening: A historical overview." *Semin Perinatol* 40 (1): 12–22.

Flake, A. W., and Harrison, M. R. 1995. "Fetal surgery." *Annu Rev Med* 46: 67–78.

Flessel, M. C., and Lorey, F. W. 2011. "The California Prenatal Screening Program: 'Options and choices,' not 'coercion and eugenics.'" *Genet Med* 13 (8): 711–13.

Gammon, B. L., Kraft, S. A., Michie, M., et al. 2016. "'I think we've got too many tests!': Prenatal providers' reflections on ethical and clinical challenges in the practice integration of cell-free DNA screening." *Ethics Med Public Health* 2 (3): 334–42.

Hartwig, T. S., Ambye, L., Sorensen, S., et al. 2017. "Discordant non-invasive prenatal testing (NIPT) — a systematic review." *Prenat Diagn* 37 (6): 527–39.

Holt, R., and Abramowicz, J. S. 2017. "Quality and safety of obstetric practices using new modalities — Ultrasound, MR, and CT." *Clin Obstet Gynecol* 60 (3): 546–61.

Jancelewicz, T., and Harrison, M. R. 2009. "A history of fetal surgery." *Clin Perinatol* 36 (2): 227–36, vii.

Johnston, J., Farrell, R. M., and Parens, E. 2017. "Supporting women's autonomy in prenatal testing." *N Engl J Med* 377 (6): 505–7.

Kaimal, A. J., Norton, M. E., and Kuppermann, M. 2015. "Prenatal testing in the genomic age: Clinical outcomes, quality of life, and costs." *Obstet Gynecol* 126 (4): 737–46.

Kuppermann, M., Norton, M. E., Thao, K., et al. 2016. "Preferences regarding contemporary prenatal genetic tests among women desiring testing: Implications for optimal testing strategies." *Prenat Diagn* 36 (5): 469–75.

Kuppermann, M., Pena, S., Bishop, J. T., et al. 2014. "Effect of enhanced information, values clarification, and removal of financial barriers on use of prenatal genetic testing: A randomized clinical trial." *JAMA* 312 (12): 1210–17.

Messerlian, G. M., and Palomaki, G. E. 2019. "Down syndrome: Overview of prenatal screening." In L. Wilkens-Haug (Ed.), *UpToDate*. Waltham, MA: Wolters Kluwer Health Division of Wolters Kluwer. https://www.uptodate.com/contents/down-syndrome-overview-of-prenatal-screening?search=Down%20syndrome:%20Overview%20of%20prenatal%20screening&source=search_result&selectedTitle=1~150&usage_type=default&display_rank=1.

Miron, P. M. 2012. "Preparation, culture, and analysis of amniotic fluid samples." *Curr Protoc Hum Genet* 74 (1): 8.4.

Moaddab, A., Nassr, A. A., Belfort, M. A., et al. 2017. "Ethical issues in fetal therapy." *Best Pract Res Clin Obstet Gynaecol* 43: 58–67.

Norton, M. E., Jacobsson, B., Swamy, G. K., et al. 2015. "Cell-free DNA analysis for noninvasive examination of trisomy." *N Engl J Med* 372 (17): 1589–97.

Palomaki, G. E., Messerlian, G. M., and Halliday, J. V. 2020. "Prenatal screening for common aneuploidies using cell-free DNA." In L. Wilkens-Haug (Ed.), *UpToDate*. Waltham, MA: Wolters Kluwer Health Division of Wolters Kluwer. https://www.uptodate.com/contents/prenatal-screening-for -common-aneuploidies-using-cell-free-dna.

Reddy, U. M., Abuhamad, A. Z., Levine, D., et al. 2014. "Fetal imaging: Executive summary of a joint Eunice Kennedy Shriver National Institute of Child Health and Human Development, Society for Maternal-Fetal Medicine, American Institute of Ultrasound in Medicine, American College of Obstetricians and Gynecologists, American College of Radiology, Society for Pediatric Radiology, and Society of Radiologists in Ultrasound fetal imaging workshop." *J Ultrasound Med* 33 (5): 745–57.

Rolfes, V., and Schmitz, D. 2016. "Unfair discrimination in prenatal aneuploidy screening using cell-free DNA?" *Eur J Obstet Gynecol Reprod Biol* 198: 27–29.

Sandlin, A. T., Ounpraseuth, S. T., Spencer, H. J., et al. 2014. "Amniotic fluid volume in normal singleton pregnancies: Modeling with quantile regression." *Arch Gynecol Obstet* 289 (5): 967–72.

Shue, E. H., Harrison, M., and Hirose, S. 2012. "Maternal-fetal surgery: History and general considerations." *Clin Perinatol* 39 (2): 269–78.

Skrzypek, H., and Hui, L. 2017. "Noninvasive prenatal testing for fetal aneuploidy and single gene disorders." *Best Pract Res Clin Obstet Gynaecol* 42: 26–38.

Society for Maternal-Fetal Medicine (SMFM) Publications Committee. 2015. SMFM Statement: Clarification of recommendations regarding cell-free DNA aneuploidy screening. *Am J Obstet Gynecol* 213 (6): 753–54. doi: 10.1016/j.ajog.2015.09.077.

Vora, N. L., Johnson, K. L., Basu, S., et al. 2012. "A multifactorial relationship exists between total circulating cell-free DNA levels and maternal BMI." *Prenat Diagn* 32 (9): 912–14.

Wax, J., Minkoff, H., Johnson, A., et al. 2014. "Consensus report on the detailed fetal anatomic ultrasound examination: Indications, components, and qualifications." *J Ultrasound Med* 33 (2): 189–95.

Witt, R., Mackenzie, T. C., and Peranteau, W. H. 2017. "Fetal stem cell and gene therapy." *Semin Fetal Neonatal Med* 22 (6): 410–14.

Wong, F. C., and Lo, Y. M. 2016. "Prenatal diagnosis innovation: Genome sequencing of maternal plasma." *Annu Rev Med* 67: 419–32.

Yaron, Y. 2016. "The implications of non-invasive prenatal testing failures: A review of an under-discussed phenomenon." *Prenat Diagn* 36 (5): 391–96.

Zhang, H., Gao, Y., Jiang, F., et al. 2015. "Non-invasive prenatal testing for trisomies 21, 18, and 13: Clinical experience from 146,958 pregnancies." *Ultrasound Obstet Gynecol* 45 (5): 530–38.

Zozzaro-Smith, P., Gray, L. M., Bacak, S. J., et al. 2014. "Limitations of aneuploidy and anomaly detection in the obese patient." *J Clin Med* 3 (3): 795–808.

CHAPTER 4

American College of Obstetricians and Gynecologists and the American Society for Reproductive Medicine. 2013. "Committee opinion no. 575: Exposure to toxic environmental agents." *Obstet Gynecol* 122 (4): 931–35. doi: 10.1097/01.AOG.0000435416.21944.54.

Aoki, Y. 2001. "Polychlorinated biphenyls, polychlorinated dibenzo-p-dioxins, and polychlorinated dibenzofurans as endocrine disrupters — What we have learned from Yusho disease." *Environ Res* 86 (1): 2–11.

Bennett, D., Bellinger, D. C., Birnbaum, L. S., et al. 2016. "Project TENDR: Targeting Environmental Neuro-Developmental Risks — The TENDR consensus statement." *Environ Health Perspect* 124 (7): A118–22. doi: 10.1289/EHP358.

Braun, J. M., and Gray, K. 2017. "Challenges to studying the health effects of early life environmental chemical exposures on children's health." *PLoS Biol* 15 (12): e2002800.

Braun, J. M., Just, A. C., Williams, P. L., et al. 2014. "Personal care product use and urinary phthalate metabolite and paraben concentrations during pregnancy among women from a fertility clinic." *J Expo Sci Environ Epidemiol* 24 (5): 459–66.

Braun, J. M., Yolton, K., Dietrich, K. N., et al. 2009. "Prenatal bisphenol A exposure and early childhood behavior." *Environ Health Perspect* 117 (12): 1945–52.

Buckley, J. P., Palmieri, R. T., Matuszewski, J. M., et al. 2012. "Consumer product exposures associated with urinary phthalate levels in pregnant women." *J Expo Sci Environ Epidemiol* 22 (5): 468–75.

Calafat, A. M., Ye, X., Wong, L. Y., et al. 2010. "Urinary concentrations of four parabens in the U.S. population: NHANES, 2005–2006." *Environ Health Perspect* 118 (5): 679–85.

Cappuccio, F. P. 2004. "Commentary: Epidemiological transition, migration, and cardiovascular disease." *Int J Epidemiol* 33 (2): 387–88.

Chen, H., Seifikar, H., Larocque, N., et al. 2019. "Using a multi-stage hESC model to characterize BDE-47 toxicity during neurogenesis." *Toxicol Sci* 171 (1): 221–34.

Choi, J., Knudsen, L. E., Mizrak, S., et al. 2017. "Identification of exposure to environmental chemicals in children and older adults using human bio-monitoring data sorted by age: Results from a literature review." *Int J Hyg Environ Health* 220 (2, Pt A): 282–98.

Collins, F. S. 2004. "The case for a US prospective cohort study of genes and environment." *Nature* 429 (6990): 475–77.

Connor, T. H., Lawson, C. C., Polovich, M., et al. 2014. "Reproductive health risks associated with occupational exposures to antineoplastic drugs in health care settings: A review of the evidence." *J Occup Environ Med* 56 (9): 901–10.

Council on Environmental Health. 2011. "Chemical-management policy: Prioritizing children's health." https://www.ncbi.nlm.nih.gov/pubmed/21518722.

Di Renzo, G. C., Conry, J. A., Blake, J., et al. 2015. "International Federation of Gynecology and Obstetrics' opinion on reproductive health impacts of exposure to toxic environmental chemicals." *Int J Gynaecol Obstet* 131 (3): 219–25.

Eskenazi, B., Chevrier, J., Rauch, S. A., et al. 2013. "In utero and childhood polybrominated diphenyl ether (PBDE) exposures and neurodevelopment in the CHAMACOS study." *Environ Health Perspect* 121 (2): 257–62.

Ezechias, M., Covino, S., and Cajthaml, T. 2014. "Ecotoxicity and biodegradability of new brominated flame retardants: A review." *Ecotoxicol Environ Saf* 110: 153–67.

Fisher, M., Macpherson, S., Braun, J. M., et al. 2017. "Paraben concentrations in maternal urine and breast milk and its association with personal care product use." *Environ Sci Technol* 51 (7): 4009–17.

Gerona, R. R., Schwartz, J. M., Pan, J., et al. 2018. "Suspect screening of maternal serum to identify new environmental chemical biomonitoring targets using liquid chromatography-quadrupole time-of-flight mass spectrometry." *J Expo Sci Environ Epidemiol* 28 (2): 101–8.

Gore, A. C., Chappell, V. A., Fenton, S. E., et al. 2015. "EDC-2: The Endocrine Society's second scientific statement on endocrine-disrupting chemicals." *Endocr Rev* 36 (6): 593–602.

Goulas, A. E., Zygoura, P., Karatapanis, A., et al. 2007. "Migration of di(2-ethylhexyl) adipate and acetyltributyl citrate plasticizers from food-grade PVC film into sweetened sesame paste (halawa tehineh): Kinetic and penetration study." *Food Chem Toxicol* 45 (4): 585–91.

Gustafson, P., Barregard, L., Strandberg, B., et al. 2007. "The impact of domestic wood burning on personal, indoor, and outdoor levels of 1,3-butadiene, benzene, formaldehyde, and acetaldehyde." *J Environ Monit* 9 (1): 23–32.

Guyton, K. Z., Loomis, D., Grosse, Y., et al. 2015. "Carcinogenicity of tetrachlorvinphos, parathion, malathion, diazinon, and glyphosate." *Lancet Oncol* 16 (5): 490–91.

Hashemipour, M., Kelishadi, R., Amin, M. M., et al. 2018. "Is there any association between phthalate exposure and precocious puberty in girls?" *Environ Sci Pollut Res Int* 25 (14): 13589–96.

Heindel, J. J., Newbold, R., and Schug, T. T. 2015. "Endocrine disruptors and obesity." *Nat Rev Endocrinol* 11 (11): 653–61.

Heindel, J. J., Skalla, L. A., Joubert, B. R., et al. 2017. "Review of developmental origins of health and disease publications in environmental epidemiology." *Reprod Toxicol* 68: 34–48.

Horan, T. S., Pulcastro, H., Lawson, C., et al. 2018. "Replacement bisphenols adversely affect mouse gametogenesis with consequences for subsequent generations." *Curr Biol* 28 (18): 2948–54 e3.

Hsu, S. T., Ma, C. I., Hsu, S. K., et al. 1985. "Discovery and epidemiology of PCB poisoning in Taiwan: A four-year followup." *Environ Health Perspect* 59: 5–10.

Kinch, C. D., Ibhazehiebo, K., Jeong, J. H., et al. 2015. "Low-dose exposure to bisphenol A and replacement bisphenol S induces precocious hypothalamic neurogenesis in embryonic zebrafish." *Proc Natl Acad Sci USA* 112 (5): 1475–80.

Krieg, S. A., Shahine, L. K., and Lathi, R. B. 2016. "Environmental exposure to endocrine-disrupting chemicals and miscarriage." *Fertil Steril* 106 (4): 941–47.

Lam, J., Lanphear, B. P., Bellinger, D., et al. 2017. "Developmental PBDE exposure and IQ/ADHD in childhood: A systematic review and meta-analysis." *Environ Health Perspect* 125 (8): 086001.

Latini, G., De Felice, C., Presta, G., et al. 2003. "In utero exposure to di-(2-eth-ylhexyl)phthalate and duration of human pregnancy." *Environ Health Perspect* 111 (14): 1783–85.

Liao, C., Liu, F., and Kannan, K. 2012. "Bisphenol s, a new bisphenol analogue, in paper products and currency bills and its association with bisphenol a residues." *Environ Sci Technol* 46 (12): 6515–22.

Lu, P. C. W., Shahbaz, S., and Winn, L. M. 2020. "Benzene and its effects on cell signaling pathways related to hematopoiesis and leukemia." *J Appl Toxicol* 40 (8): 1018–32.

Magriplis, E., Farajian, P., Panagiotakos, D. B., et al. 2017. "Maternal smoking and risk of obesity in schoolchildren: Investigating early life theory from the GRECO study." *Prev Med Rep* 8: 177–82.

McNamara, P. J, and Levy, S. B. 2016. "Triclosan: An instructive tale." *Antimicrob Agents Chemother* 60 (12): 7015–16.

Mocarelli, P., Gerthoux, P. M., Needham, L. L., et al. 2011. "Perinatal exposure to low doses of dioxin can permanently impair human semen quality." *Environ Health Perspect* 119 (5): 713–18.

Morello-Frosch, R., Cushing, L. J., Jesdale, B. M., et al. 2016. "Environmental chemicals in an urban population of pregnant women and their newborns from San Francisco." *Environ Sci Technol* 50 (22): 12464–72.

Mrema, E. J., Rubino, F. M., Brambilla, G., et al. 2013. "Persistent organochlorinated pesticides and mechanisms of their toxicity." *Toxicology* 307: 74–88.

Newbold, R. R. 2011. "Developmental exposure to endocrine-disrupting chemicals programs for reproductive tract alterations and obesity later in life." *Am J Clin Nutr* 94 (6 Suppl): 1939S–1942S.

Parry, E., Zota, A. R., Park, J. S., and Woodruff, T. J. 2018. "Polybrominated diphenyl ethers (PBDEs) and hydroxylated PBDE metabolites (OH-PB-DEs): A six-year temporal trend in Northern California pregnant women." *Chemosphere* 195: 777–83.

Peto, J. 2001. "Cancer epidemiology in the last century and the next decade." *Nature* 411 (6835): 390–95.

Prüss-Ustün, A., Vickers, C., Haefliger, P., et al. 2011. "Knowns and unknowns on burden of disease due to chemicals: A systematic review." *Environ Health* 10: 9.

Rappaport, S. M., and Smith, M. T. 2010. "Epidemiology: Environment and disease risks." *Science* 330 (6003): 460–61.

Rich, D. Q., Liu, K., Zhang, J., et al. 2015. "Differences in birth weight associated with the 2008 Beijing Olympics air pollution reduction: Results from a natural experiment." *Environ Health Perspect* 123 (9): 880–87.

Robinson, J. F., Kapidzic, M., Hamilton, E. G., et al. 2019. "Genomic profiling of BDE-47 effects on human placental cytotrophoblasts." *Toxicol Sci* 167 (1): 211–26.

Shelton, J. F., Geraghty, E. M., Tancredi, D. J., et al. 2014. "Neurodevelopmental disorders and prenatal residential proximity to agricultural pesticides: The CHARGE study." *Environ Health Perspect* 122 (10): 1103–9.

Shirangi, A., Fritschi, L., and Holman, C. D. 2008. "Maternal occupational exposures and risk of spontaneous abortion in veterinary practice." *Occup Environ Med* 65 (11): 719–25.

Smith, K. W., Braun, J. M., Williams, P. L., et al. 2012. "Predictors and variability of urinary paraben concentrations in men and women, including before and during pregnancy." *Environ Health Perspect* 120 (11): 1538–43.

Stotland, N. E., Sutton, P., Trowbridge, J., et al. 2014. "Counseling patients on preventing prenatal environmental exposures — A mixed-methods study of obstetricians." *PLoS One* 9 (6): e98771.

Sugiura-Ogasawara, M., Ozaki, Y., Sonta, S., et al. 2005. "Exposure to bisphenol A is associated with recurrent miscarriage." *Hum Reprod* 20 (8): 2325–29.

Sutton, P., Woodruff, T. J., Perron, J., et al. 2012. "Toxic environmental chemicals: The role of reproductive health professionals in preventing harmful exposures." *Am J Obstet Gynecol* 207 (3): 164–73.

Swan, S. H., Liu, F., Hines, M., et al. 2010. "Prenatal phthalate exposure and reduced masculine play in boys." *Int J Androl* 33 (2): 259–69.

Swan, S. H., Main, K. M., Liu, F., et al. 2005. "Decrease in anogenital distance among male infants with prenatal phthalate exposure." *Environ Health Perspect* 113 (8): 1056–61.

US Department of Health and Human Services. 2009. "Fourth national report on human exposure to environmental chemicals." *CDC Executive Summary* (updated 2019). https://search.cdc.gov/search/?query=Fourth+National+Report+on+Human+Exposure+to+Environmental+Chemicals&utf8=%E2%9C%93&affiliate=cdc-main#content.

Varshavsky, J. R., Morello-Frosch, R., Woodruff, T. J., et al. 2018. "Dietary sources of cumulative phthalates exposure among the US general population in NHANES 2005–2014." *Environ Int* 115: 417–29.

Varshavsky, J. R., Zota, A. R., and Woodruff, T. J. 2016. "A novel method for calculating potency-weighted cumulative phthalates exposure with implications for identifying racial/ethnic disparities among US reproductive-aged women in NHANES 2001–2012." *Environ Sci Technol* 50 (19): 10616–24.

Vogel, S. A. 2009. "The politics of plastics: The making and unmaking of bisphenol A 'safety.'" *Am J Public Health* 99 (Suppl 3): S559–S566.

Wald, N. J. 2004. "Folic acid and the prevention of neural-tube defects." *N Engl J Med* 350 (2): 101–3.

Wang, A., Padula, A., Sirota, M., et al. 2016. "Environmental influences on reproductive health: The importance of chemical exposures." *Fertil Steril* 106 (4): 905–29.

Washino, N., Saijo, Y., Sasaki, S., et al. 2009. "Correlations between prenatal exposure to perfluorinated chemicals and reduced fetal growth." *Environ Health Perspect* 117 (4): 660–67.

Weatherly, L. M., and Gosse, J. A. 2017. "Triclosan exposure, transformation, and human health effects." *J Toxicol Environ Health B Crit Rev* 20 (8): 447–69.

Wigle, D. T., Turner, M. C., and Krewski, D. 2009. "A systematic review and meta-analysis of childhood leukemia and parental occupational pesticide exposure." *Environ Health Perspect* 117 (10): 1505–13.

Wild, C. P. 2005. "Complementing the genome with an "exposome": The outstanding challenge of environmental exposure measurement in molecular epidemiology." *Cancer Epidemiol Biomarkers Prev* 14 (8): 1847–50.

Woodruff, T. J., Zota, A. R., and Schwartz, J. M. 2011. "Environmental chemicals in pregnant women in the United States: NHANES 2003–2004." *Environ Health Perspect* 119 (6): 878–85.

Yorifuji, T., Kato, T., Kado, Y., et al. 2015. "Intrauterine exposure to methylmercury and neurocognitive functions: Minamata disease." *Arch Environ Occup Health* 70 (5): 297–302.

Zlatnik, M. G. 2016. "Endocrine-disrupting chemicals and reproductive health." *J Midwifery Womens Health* 61 (4): 442–55.

Zota, A. R., Linderholm, L., Park, J. S., et al. 2013. "Temporal comparison of PBDEs, OH-PBDEs, PCBs, and OH-PCBs in the serum of second trimester pregnant women recruited from San Francisco General Hospital, California." *Environ Sci Technol* 47 (20): 11776–84.

Zota, A. R., and Shamasunder, B. 2017. "The environmental injustice of beauty: Framing chemical exposures from beauty products as a health disparities concern." *Am J Obstet Gynecol* 217 (4): 418e1–418e6.

CHAPTER 5

Alfirevic, Z., Stampalija, T., and Medley, N. 2017. "Cervical stitch (cerclage) for preventing preterm birth in singleton pregnancy." *Cochrane Database Syst Rev* 6 (6): CD008991. doi: 10.1002/14651858.CD008991.pub3.

American College of Obstetricians and Gynecologists. 2013. "Report of the American College of Obstetricians and Gynecologists' Task Force on Hypertension in Pregnancy." *Obstet Gynecol* 122 (5): 1122–31. doi: 10.1097/01. AOG.0000437382.03963.88.

——— 2016. "Practice advisory on low-dose aspirin and prevention of preeclampsia: Updated recommendations." http://www.losolivos-obgyn.com/info/md/acog/Low-dose%20aspirin,%20ACOG%20Practice%20Advisory%202016.pdf.

——— 2017. Frequently asked questions: Pregnancy — Gestational diabetes," FAQ177. https://www.acog.org/patient-resources/faqs/pregnancy/gestational-diabetes.

——— 2018. Committee on Practice Bulletins — Obstetrics. "Practice bulletin no. 190: Gestational diabetes mellitus." *Obstet Gynecol* 131 (2): e49–e64. doi: 10.1097/AOG.0000000000002501.

——— 2019. "Practice bulletin no. 202: "Gestational Hypertension and Preeclampsia." *Obstet Gynecol* 133 (1): e1–e25. doi: 10.1097/AOG.0000000000003018.

——— 2015. "Committee opinion no. 638: First-trimester risk assessment for early-onset preeclampsia." *Obstet Gynecol* 126 (3): e25–e27. doi: 10.1097/AOG.0000000000001049.

——— 2015. "Committee opinion no. 638: First-trimester risk assessment for early-onset preeclampsia." *Obstet Gynecol* 126 (3). 2016. "Committee opinion no. 652: Magnesium sulfate use in obstetrics." *Obstet Gynecol* 127 (1): e52–e53. doi: 10.1097/AOG.0000000000001267.

Baeyens, L., Hindi, S., Sorenson, R. L., et al. 2016. "Beta-cell adaptation in pregnancy." *Diabetes Obes Metab* 18 (Suppl 1): 63–70.

Barros, F. C., Bhutta, Z. A., Batra, M., et al. 2010. "Global report on preterm birth and stillbirth (3 of 7): Evidence for effectiveness of interventions." *BMC Pregnancy Childbirth* 10 (Suppl 1): S3.

Bellamy, L., Casas, J. P., Hingorani, A. D., et al. 2007. "Pre-eclampsia and risk of cardiovascular disease and cancer in later life: Systematic review and meta-analysis." *BMJ* 335 (7627): 974.

Bigelow, C. A., Pereira, G. A., Warmsley, A., et al. 2014. "Risk factors for new-onset late postpartum preeclampsia in women without a history of preeclampsia." *Am J Obstet Gynecol* 210 (4): 338e1–338e8.

Borchers, A. T., Naguwa, S. M., Keen, C. L., et al. 2010. "The implications of autoimmunity and pregnancy." *J Autoimmun* 34 (3): J287–J299.

Burton, G. J., Redman, C. W., Roberts, J. M., et al. 2019. "Pre-eclampsia: Pathophysiology and clinical implications." *BMJ* 366: l2381.

Callaghan, W. M., Creanga, A. A., and Kuklina, E. V. 2012. "Severe maternal morbidity among delivery and postpartum hospitalizations in the United States." *Obstet Gynecol* 120 (5): 1029–36.

Centers for Disease Control and Prevention. 2017. "Data on selected pregnancy complications in the United States." https://www.cdc.gov/reproductivehealth/maternalinfanthealth/pregnancy-complications-data.htm.

Cha, J. M., and Aronoff, D. M. 2017. "A role for cellular senescence in birth timing." *Cell Cycle* 16 (21): 2023–31.

Chappell, L. C., Brocklehurst, P., Green, M. E., et al. 2019. "Planned early delivery or expectant management for late preterm pre-eclampsia (PHOENIX): A randomised controlled trial." *Lancet* 394 (10204): 1181–90.

Davenport, M. H., Ruchat, S. M., Poitras, V. J., et al. 2018. "Prenatal exercise for the prevention of gestational diabetes mellitus and hypertensive disorders of pregnancy: A systematic review and meta-analysis." *Br J Sports Med* 52 (21): 1367–75.

Durnwald, C. 2020. "Diabetes mellitus in pregnancy: Screening and diagnosis." In D. M. Nathan and E. F. Werner (Eds.), *UpToDate*. Waltham, MA: Wolters Kluwer Health Division of Wolters Kluwer. https://www.uptodate.com/contents/diabetes-mellitus-in-pregnancy-screening-and-diagnosis#H132644621.

Fisher, S. J. 2015. "Why is placentation abnormal in preeclampsia?" *Am J Obstet Gynecol* 213 (4 Suppl): S115–S1122.

Goldenberg, R. L. 2002. "The management of preterm labor." *Obstet Gynecol* 100 (5, Pt 1): 1020–37.

Greene, M. F., and Solomon, C. G. 2017. "Aspirin to prevent preeclampsia." *N Engl J Med* 377 (7): 690–91.

Gyamfi-Bannerman, C., Thom, E. A., Blackwell, S. C., et al. 2016. "Antenatal betamethasone for women at risk for late preterm delivery." *N Engl J Med* 374 (14): 1311–20.

Haas, D. M., Morgan, A. M., Deans, S. J., et al. 2015. "Ethanol for preventing preterm birth in threatened preterm labor." *Cochrane Database Syst Rev* 11: CD011445. doi: 10.1002/14651858.CD011445.pub2.

Harville, E. , Xiong, X., and Buekens, P. 2010. "Disasters and perinatal health: A systematic review." *Obstet Gynecol Surv* 65 (11): 713–28.

Henderson, J. T., Thompson, J. H., Burda, B. U., et al. 2017. "Screening for preeclampsia: A systematic evidence review for the U.S. Preventive Services Task Force. Rockville, MD. https://www.ncbi.nlm.nih.gov/books/NBK447462.

Hernandez, T. L., Mande, A., and Barbour, L. A. 2018. "Nutrition therapy within and beyond gestational diabetes." *Diabetes Res Clin Pract* 145: 39–50.

Hofmeyr, G. J., Lawrie, T. A., Atallah, A. N., et al. 2014. "Calcium supplementation during pregnancy for preventing hypertensive disorders and related problems." *Cochrane Database Syst Rev* CD001059. doi: 10.1002/14651858.CD001059.pub4.

Kaaja, R. J., and Greer, I. A. 2005. "Manifestations of chronic disease during pregnancy." *JAMA* 294 (21): 2751–57.

Karumanchi, S. A., and Granger, J. P. 2016. "Preeclampsia and pregnancy-related hypertensive disorders." *Hypertension* 67 (2): 238–42.

Lee, Y. M., Kim, S. A., Lee, I. K., et al. 2016. "Effect of a brown rice based vegan diet and conventional diabetic diet on glycemic control of patients with type 2 diabetes: A 12-week randomized clinical trial." *PLoS One* 11 (6): e0155918.

Lisonkova, S., and Joseph, K. S. 2013. "Incidence of preeclampsia: Risk factors and outcomes associated with early- versus late-onset disease." *Am J Obstet Gynecol* 209 (6): 544e1–544e12.

Lo, C. C. W., Lo, A. C. Q., Leow, S. H., et al. 2020. "Future cardiovascular disease risk for women with gestational hypertension: A systematic review and meta-analysis." *J Am Heart Assoc*: e013991.

March of Dimes. 2018. "Premature birth report cards." https://www.marchofdimes.org/materials/PrematureBirthReportCard-United%20States-2018.pdf.

Melvin, L. M., and Funai, E. F. 2020. "Gestational hypertension." In C. J. Lockwood (Ed.), *UpToDate*. Waltham, MA: Wolters Kluwer Health Division of Wolters Kluwer. https://www.uptodate.com/contents/gestational

-hypertension?search=gestational%20hypertension&source=search_result &selectedTitle=1~94&usage_type=default&display_rank=1.

Metzger, B. E., Lowe, L. P., Dyer, A. R., et al. (Hapo Study Cooperative Research Group.) 2008. "Hyperglycemia and adverse pregnancy outcomes." *N Engl J Med* 358 (19): 1991–2002.

Molina, R. L., and Pace, L. E. 2017. "A renewed focus on maternal health in the United States." *N Engl J Med* 377 (18): 1705–7.

Muglia, L. J., and Katz, M. 2010. "The enigma of spontaneous preterm birth." *N Engl J Med* 362 (6): 529–35.

Mustafa, R., Ahmed, S., Gupta, A., et al. 2012. "A comprehensive review of hypertension in pregnancy." *J Pregnancy* 2012: 105918.

Norton, M. 2012. "Serious birth complications rising in the U.S." Reuters.

Norwitz, E. R. 2020. "Progesterone supplementation to reduce the risk of spontaneous preterm birth." In C. J. Lockwood (Ed.), *UpToDate*. Waltham, MA: Wolters Kluwer Health Divison of Wolters Kluwer. https://www .uptodate.com/contents/progesterone-supplementation-to-reduce-the-risk -of-spontaneous-preterm-birth?search=progesterone%20supplementation %20to%20reduce%20the%20risk%20of%20spontaneous%20preterm %20birth&source=search_result&selectedTitle=1~150&usage_type=default &display_rank=1.

Norwitz, E. R., and Caughey, A. B. 2011. "Progesterone supplementation and the prevention of preterm birth." *Rev Obstet Gynecol* 4 (2): 60–72. https:// www.ncbi.nlm.nih.gov/pubmed/22102929.

Plunkett, J., Feitosa, M. F., Trusgnich, M., et al. 2009. "Mother's genome or maternally-inherited genes acting in the fetus influence gestational age in familial preterm birth." *Hum Hered* 68 (3): 209–19.

Ramsey, P. S., and Schenken, R. S. 2020. "Obesity in pregnancy: Complications and maternal management." In C. J. Lockwood and F. X. Pi-Sunyer (Eds.), *UpToDate*. Waltham, MA: Wolters Kluwer Health Division of Wolters Kluwer. https://www.uptodate.com/contents/obesity-in -pregnancy-complications-and-maternal-management?search=obesity %20in%20pregnancy:%20Complications%20and%20maternal&source= search_result&selectedTitle=1~150&usage_type=default&display _rank=1.

Rana, S., Lemoine, E., Granger, J. P., et al. 2019. "Preeclampsia: Pathophysiology, challenges, and perspectives." *Circ Res* 124 (7): 1094–12.

Redman, C. W., and Sargent, I. L. 2005. "Latest advances in understanding preeclampsia." *Science* 308 (5728): 1592–94.

Roberge, S., Bujold, E., and Nicolaides, K. H. 2018. "Aspirin for the prevention of preterm and term preeclampsia: Systematic review and metaanalysis." *Am J Obstet Gynecol* 218 (3): 287–93e1.

Roberts, J. M. 1998. "Endothelial dysfunction in preeclampsia." *Semin Reprod Endocrinol* 16 (1): 5–15.

Rolnik, D. L., Wright, D. , Poon, L. C., et al. 2017. "Aspirin versus placebo in pregnancies at high risk for preterm preeclampsia." *N Engl J Med* 377 (7): 613–22.

Romero, R., Dey, S. K., and Fisher, S. J. 2014. "Preterm labor: One syndrome, many causes." *Science* 345 (6198): 760–65.

Rumbold, A., Duley, L., Crowther, C. A., et al. 2008. "Antioxidants for preventing pre-eclampsia." *Cochrane Database Syst Rev*, CD004227. doi: 10.1002/14651858.CD004227.pub3.

San Francisco Department of Public Health. 2014. "Preterm birth: Indicators, causes, and prevention strategies in San Francisco." https://www.sfdph.org/dph/hc/HCCommPublHlth/Agendas/2014/November%2018/Preterm%20Births_Fact%20Sheet_SFDPH_v20140930k.pdf.

Sircar, M., Thadhani, R., and Karumanchi, S. A. 2015. "Pathogenesis of preeclampsia." *Curr Opin Nephrol Hypertens* 24 (2): 131–38.

Society for Maternal-Fetal Medicine. 2013. "Evaluation and management of severe preeclampsia." https://www.smfm.org/publications/91-evaluation-and-management-of-severe-preeclampsia.

Sosa, C. G., Althabe, F., Belizan, J. M., et al. 2015. "Bed rest in singleton pregnancies for preventing preterm birth." *Cochrane Database Syst Rev* CD003581. doi: 10.1002/14651858.CD003581.pub3.

Soto-Wright, V., Bernstein, M., Goldstein, D. P., et al. 1995. "The changing clinical presentation of complete molar pregnancy." *Obstet Gynecol* 86 (5): 775–79.

Story, L., and Chappell, L. C. 2017. "Preterm pre-eclampsia: What every neonatologist should know." *Early Hum Dev* 114: 26–30.

Thompson, S. V., Winham, D. M., and Hutchins, A. M. 2012. "Bean and rice meals reduce postprandial glycemic response in adults with type 2 diabetes: A cross-over study." *Nutr J* 11: 23. doi: 10.1186/1475-2891-11-23.

US Preventive Services Task Force, Bibbins-Domingo, K., Grossman, D. C., et al. 2017. "Screening for preeclampsia: US Preventive Services Task Force recommendation statement. *JAMA* 317 (16): 1661–67. doi: 10.1001/jama.2017.3439.

York, T. P., Eaves, L. J., Lichtenstein, P., et al. 2013. "Fetal and maternal genes' influence on gestational age in a quantitative genetic analysis of 244,000 Swedish births." *Am J Epidemiol* 178 (4): 543–50.

Zhang, G., Feenstra, B., Bacelis, J., et al. 2017. "Genetic associations with gestational duration and spontaneous preterm birth." *N Engl J Med* 377 (12): 1156–67.

Zhou, Y., Mcmaster, M., Woo, K., et al. 2002. "Vascular endothelial growth factor ligands and receptors that regulate human cytotrophoblast survival are dysregulated in severe preeclampsia and hemolysis, elevated liver enzymes, and low platelets syndrome." *Am J Pathol* 160 (4): 1405–23.

CHAPTER 6

Agostoni, C., and Fattore, G. 2013. "Growth outcome: Nutritionist perspective." *World Rev Nutr Diet* 106: 12–8.

Alexander, B. T., Dasinger, J. H., and Intapad, S. 2015. "Fetal programming and cardiovascular pathology." *Compr Physiol* 5 (2): 997–1025.

American College of Obstreticians and Gynecologists. 2020. "Committee opinion no. 804: Physical activity and exercise during pregnancy and the postpartum period." *Obstet Gynecol* 135: e178–e188. doi: 10.1097/AOG.0000000000003772.

Armitage, J. A., Poston, L., and Taylor, P. D. 2008. "Developmental origins of obesity and the metabolic syndrome: The role of maternal obesity." *Front Horm Res* 36: 73–84.

Barker, D. J. 1993. "Fetal origins of coronary heart disease." *Br Heart J* 69 (3): 195–96.

Barker, D. J., and Osmond, C. 1986. "Infant mortality, childhood nutrition, and ischaemic heart disease in England and Wales." *Lancet* 1 (8489): 1077–81.

Barker, D. J. , Osmond, C., Forsen, T. J., et al. 2005. "Trajectories of growth among children who have coronary events as adults." *N Engl J Med* 353 (17): 1802–9.

——. 2007. "Maternal and social origins of hypertension." *Hypertension* 50 (3): 565–71.

Barker, D. J., Osmond, C., Golding, J., et al. 1989. "Growth in utero, blood pressure in childhood and adult life, and mortality from cardiovascular disease." *BMJ* 298 (6673): 564–67.

Barker, D. J., Osmond, C., Kajantie, E., et al. 2009. "Growth and chronic disease: Findings in the Helsinki Birth Cohort." *Ann Hum Biol* 36 (5): 445–58.

Barker, D. J., and Thornburg, K. L. 2013. "The obstetric origins of health for a lifetime." *Clin Obstet Gynecol* 56 (3): 511–19.

———. 2013. "Placental programming of chronic diseases, cancer, and lifespan: A review." *Placenta* 34 (10): 841–45.

Barker, D. J., Thornburg, K. L., Osmond, C., et al. 2010. "The surface area of the placenta and hypertension in the offspring in later life." *Int J Dev Biol* 54 (2–3): 525–30.

Barker, D. J., Winter, P. D., Osmond, C., et al. 1989. "Weight in infancy and death from ischaemic heart disease." *Lancet* 2 (8663): 577–80.

Bavdekar, A., Yajnik, C. S., Fall, C. H., et al. 1999. "Insulin resistance syndrome in 8-year-old Indian children: Small at birth, big at 8 years, or both?" *Diabetes* 48 (12): 2422–29.

Bellatorre, A., Scherzinger, A., Stamm, E., et al. 2018. "Fetal overnutrition and adolescent hepatic fat fraction: The Exploring Perinatal Outcomes in Children study." *J Pediatr* 192: 165–70 e1.

Bhargava, S. K., Sachdev, H. S., Fall, C. H., et al. 2004. "Relation of serial changes in childhood body-mass index to impaired glucose tolerance in young adulthood." *N Engl J Med* 350 (9): 865–75.

Bianco-Miotto, T., Craig, J. M., Gasser, Y. P., et al. 2017. "Epigenetics and DOHaD: From basics to birth and beyond." *J Dev Orig Health Dis* 8 (5): 513–19.

Birnbaum, L. S., and Miller, M. F. 2015. "Prenatal programming and toxicity (PPTOX) introduction." *Endocrinology* 156 (10): 3405–7.

Bisson, M., Almeras, N., Plaisance, J., et al. 2013. "Maternal fitness at the onset of the second trimester of pregnancy: Correlates and relationship with infant birth weight." *Pediatr Obes* 8 (6): 464–74.

Bjorn, G. 2008. "As obesity epidemic grows, research shows fitness benefits fetal development." *Nature Medicine* 14: 1167.

Blanton, L. V., Barratt, M. J., Charbonneau, M. R., et al. 2016. "Childhood undernutrition, the gut microbiota, and microbiota-directed therapeutics." *Science* 352 (6293): 1533.

Blanton, L. V., Charbonneau, M. R., Salih, T., et al. 2016. "Gut bacteria that prevent growth impairments transmitted by microbiota from malnourished children." *Science* 351 (6275): 830.

Boney, C. M., Verma, A., Tucker, R., et al. 2005. "Metabolic syndrome in childhood: Association with birth weight, maternal obesity, and gestational diabetes mellitus." *Pediatrics* 115 (3): e290–6.

Burton, G. J., Fowden, A. L., and Thornburg, K. L. 2016. "Placental origins of chronic disease." *Physiol Rev* 96 (4): 1509–65.

Castillo-Fernandez, J. E., Spector, T. D., and Bell, J. T. 2014. "Epigenetics of discordant monozygotic twins: Implications for disease." *Genome Med* 6 (7): 60.

Chan, J., Natekar, A., and Koren, G. 2014. "Hot yoga and pregnancy: Fitness and hyperthermia." *Can Fam Physician* 60 (1): 41–42.

Chapin, H. D. 1909. "Biology as the basic principle in infant feeding." *The Post Grad* 23 (3): 272–80.

Charbonneau, M. R, Blanton, L. V., Digiulio, D. B., et al. 2016. "A microbial perspective of human developmental biology." *Nature* 535 (7610): 48–55.

Chavatte-Palmer, P., Tarrade, A., and Rousseau-Ralliard, D. 2016. "Diet before and during pregnancy and offspring health: The importance of animal models and what can be learned from them." *Int J Environ Res Public Health* 13 (6).

Clapp, J. F. 3rd. 2006. "Influence of endurance exercise and diet on human placental development and fetal growth." *Placenta* 27 (6–7): 527–34.

——. 2008. "Long-term outcome after exercising throughout pregnancy: Fitness and cardiovascular risk." *Am J Obstet Gynecol* 199 (5): 489 e1–e6.

——. 2009. "Is exercise during pregnancy related to preterm birth?" *Clin J Sport Med* 19 (3): 241–43.

Clapp, J. F. 3rd, Kim, H., Burciu, B., et al. 2000. "Beginning regular exercise in early pregnancy: Effect on fetoplacental growth." *Am J Obstet Gynecol* 183 (6): 1484–88.

Cooper, C. 2013. "David Barker (1938–2013)." *Nature* 502 (7471): 304.

Couzin-Frankel, J. 2013. "Mysteries of development: How does fetal environment influence later health?" *Science* 340 (6137): 1160–61.

Dabelea, D., Hanson, R. L., Lindsay, R. S., et al. 2000. "Intrauterine exposure to diabetes conveys risks for type 2 diabetes and obesity: A study of discordant sibships." *Diabetes* 49 (12): 2208–11.

Desai, M., and Cot, M. 2015. "Epidemiology of malaria during pregnancy: Burden and impact of Plasmodium falciparum malaria on maternal infant health." *Encyclopedia of Malaria*: 1–13.

Di Mascio, D., Magro-Malosso, E. R., Saccone, G., et al. 2016. "Exercise during pregnancy in normal-weight women and risk of preterm birth: A systematic review and meta-analysis of randomized controlled trials." *Am J Obstet Gynecol* 215 (5): 561–71.

Engel, S. M., Berkowitz, G. S., Wolff, M. S., et al. 2005. "Psychological trauma associated with the World Trade Center attacks and its effect on pregnancy outcome." *Paediatr Perinat Epidemiol* 19 (5): 334–41.

Eriksson, J. G. 2011. "Early growth and coronary heart disease and type 2 diabetes: Findings from the Helsinki Birth Cohort Study (HBCS)." *Am J Clin Nutr* 94 (6 Suppl): 1799S–1802S.

Eriksson, J., Forsen, T., Tuomilehto, J., et al. 2000. "Fetal and childhood growth and hypertension in adult life." *Hypertension* 36 (5): 790–94.

———. 2003. "Early adiposity rebound in childhood and risk of type 2 diabetes in adult life." *Diabetologia* 46 (2): 190–94.

Ernst, P. B., and Gold, B. D. 2000. "The disease spectrum of *Helicobacter pylori:* The immunopathogenesis of gastroduodenal ulcer and gastric cancer." *Annu Rev Microbiol* 54: 615–40.

Fall, C. H. 2013. "Fetal programming and the risk of noncommunicable disease." *Indian J Pediatr* 80 (Suppl 1): S13–S20.

Fall, C. H., Barker, D. J., Osmond, C., et al. 1992. "Relation of infant feeding to adult serum cholesterol concentration and death from ischaemic heart disease." *BMJ* 304 (6830): 801–5.

Feig, D. S., Lipscombe, L. L, Tomlinson, G., et al. 2011. "Breastfeeding predicts the risk of childhood obesity in a multi-ethnic cohort of women with diabetes." *J Matern Fetal Neonatal Med* 24 (3): 511–15.

Hachey, D. L. 1994. "Benefits and risks of modifying maternal fat intake in pregnancy and lactation." *Am J Clin Nutr* 59 (2 Suppl): 454S–463S.

Hales, C. N., and Barker, D. J. 1992. "Type 2 (non-insulin-dependent) diabetes mellitus: The thrifty phenotype hypothesis." *Diabetologia* 35 (7): 595–601.

———. 2001. "The thrifty phenotype hypothesis." *Br Med Bull* 60: 5–20.

———. 2013. "Type 2 (non-insulin-dependent) diabetes mellitus: The thrifty phenotype hypothesis. 1992." *Int J Epidemiol* 42 (5): 1215–22.

Hall, S. S. 2007, Nov. 12. "Small and thin: The controversy over the fetal origins of adult health." *New Yorker*: 52–57.

Harder, T., Bergmann, R., Kallischnigg, G., et al. 2005. "Duration of breast-feeding and risk of overweight: A meta-analysis." *Am J Epidemiol* 162 (5): 397–403.

Heijmans, B. T., Tobi, E. W., Stein, A. D., et al. 2008. "Persistent epigenetic differences associated with prenatal exposure to famine in humans." *Proc Natl Acad Sci USA* 105 (44): 17046–49.

Hochner, H., Friedlander, Y., Calderon-Margalit, R., et al. 2012. "Associations of maternal prepregnancy body mass index and gestational weight gain with adult offspring cardiometabolic risk factors: The Jerusalem perinatal family follow-up study." *Circulation* 125 (11): 1381–89.

Hohwu, L., Li, J., Olsen, J., et al. 2014. "Severe maternal stress exposure due to bereavement before, during, and after pregnancy and risk of overweight and obesity in young adult men: A Danish national cohort study." *PLoS One* 9 (5): e97490.

Hopkins, S. A., Baldi, J. C., Cutfield, W. S., et al. 2010. "Exercise training in pregnancy reduces offspring size without changes in maternal insulin sensitivity." *J Clin Endocrinol Metab* 95 (5): 2080–88.

Horta, B. L., Loret De Mola, C., and Victora, C. G. 2015. "Long-term consequences of breastfeeding on cholesterol, obesity, systolic blood pressure, and type 2 diabetes: A systematic review and meta-analysis." *Acta Paediatr* 104 (467): 30–37.

Jackson, M. R., Gott, P., Lye, S. J., Ritchie, J. W., and Clapp, J. F., 3rd. 1995. "The effects of maternal aerobic exercise on human placental development: Placental volumetric composition and surface areas." *Placenta* 16 (2): 179–91.

Kalliomaki, M., Collado, M. C., Salminen, S., et al. 2008. "Early differences in fecal microbiota composition in children may predict overweight." *Am J Clin Nutr* 87 (3): 534–38.

Kiserud, T., Piaggio, G., Carroli, G., et al. 2017. "The World Health Organization fetal growth charts: A multinational longitudinal study of ultrasound biometric measurements and estimated fetal weight." *PLoS Med* 14 (1): e1002220.

Knight, B. A., Shields, B. M., Brook, A., et al. 2015. "Lower circulating B12 is associated with higher obesity and insulin resistance during pregnancy in a non-diabetic white British population." *PLoS One* 10 (8): e0135268.

Krishnaveni, G. V., Hill, J. C., Veena, S. R., et al. 2009. "Low plasma vitamin

B12 in pregnancy is associated with gestational 'diabesity' and later diabetes." *Diabetologia* 52 (11):2350–8.

Latal-Hajnal, B., Von Siebenthal, K., Kovari, H., et al. 2003. "Postnatal growth
in VLBW infants: Significant association with neurodevelopmental outcome." *J Pediatr* 143 (2): 163–70.

Li, Y., Jaddoe, V. W., Qi, L., et al. 2011. "Exposure to the Chinese famine in
early life and the risk of hypertension in adulthood." *J Hypertens* 29 (6):
1085–92.

Liggins Institute, TUOA. "IMPROVE trial: Improving maternal and progeny
risk of obesity via exercise." http://www.liggins.auckland.ac.nz/en/about/res
earchtranslationandfacilities/clinicalresea.

Lim, R., and Sobey, C. G. 2011. "Maternal nicotine exposure and fetal programming of vascular oxidative stress in adult offspring." *Br J Pharmacol* 164
(5): 1397–99.

The Low Birth Weight and Nephron Number Working Group. 2017. "The
impact of kidney development on the life course: A consensus document for
action." *Nephron* 136 (1): 3–49.

Lumey, L. H., Stein, A. D., and Susser, E. 2011. "Prenatal famine and adult
health." *Annu Rev Public Health* 32: 237–62.

Mahajan, A., Sapehia, D., Thakur, S., et al. 2019. "Effect of imbalance in folate
and vitamin B12 in maternal/parental diet on global methylation and regulatory miRNAs." *Sci Rep* 9 (1): 17602.

Marmot, M. G., Page, C. M., Atkins, E., et al. 1980. "Effect of breast-feeding on
plasma cholesterol and weight in young adults." *J Epidemiol Community
Health* 34 (3): 164–67.

Mericq, V., Martinez-Aguayo, A., Uauy, R., et al. 2017. "Long-term metabolic
risk among children born premature or small for gestational age." *Nat Rev
Endocrinol* 13 (1): 50–62.

Norris, T., McCarthy, F. P., Khashan, A. S., et al. 2017. "Do changing levels of
maternal exercise during pregnancy affect neonatal adiposity? Secondary
analysis of the babies after SCOPE: Evaluating the longitudinal impact
using neurological and nutritional endpoints (BASELINE) birth cohort
(Cork, Ireland)." *BMJ Open* 7 (11): e017987.

O'Connor, P. J., Poudevigne, M. S., Cress, M. E., et al. 2011. "Safety and efficacy
of supervised strength training adopted in pregnancy." *J Phys Act Health* 8
(3): 309–20.

Otto, M. W., Church, T. S., Craft, L. L., et al. 2007. "Exercise for mood and anxiety disorders." *Prim Care Companion J Clin Psychiatry* 9 (4): 287–94.

Pannaraj, P. S., Li, F., Cerini, C., et al. 2017. "Association between breast milk bacterial communities and establishment and development of the infant gut microbiome." *JAMA Pediatr* 171 (7): 647–54.

Papageorghiou, A. T., Ohuma, E. O., Altman, D. G., et al. 2014. "International standards for fetal growth based on serial ultrasound measurements: The Fetal Growth Longitudinal Study of the INTERGROWTH-21st Project." *Lancet* 384 (9946): 869–79.

Pennisi, E. 2013. "How do microbes shape animal development?" *Science* 340 (6137): 1159–60.

———. 2016. "Microbiome: The right gut microbes help infants grow." *Science* 351 (6275): 802.

Pirkola, J., Pouta, A., Bloigu, A., et al. 2010. "Risks of overweight and abdominal obesity at age 16 years associated with prenatal exposures to maternal prepregnancy overweight and gestational diabetes mellitus." *Diabetes Care* 33 (5): 1115–21.

Razani, N., Kohn, M. A., Wells, N. M., et al. 2016. "Design and evaluation of a park prescription program for stress reduction and health promotion in low-income families: The Stay Healthy in Nature Everyday (SHINE) study protocol." *Contemp Clin Trials* 51: 8–14.

Roseboom, T., De Rooij, S., and Painter, R. 2006. "The Dutch famine and its long-term consequences for adult health." *Early Hum Dev* 82 (8): 485–91.

Rosenfeld, C. S. 2017. "Homage to the 'H' in developmental origins of health and disease." *J Dev Orig Health Dis* 8 (1): 8–29.

Scharschmidt, T. C. 2017. "Growing good bugs with mom's milk." *Science Translational Medicine* 9 (412): eaap8172.

Susser, E., St. Clair, D., and He, L. 2008. "Latent effects of prenatal malnutrition on adult health: The example of schizophrenia." *Ann N Y Acad Sci* 1136: 185–92.

Suzuki, K. 2018. "The developing world of DOHaD." *J Dev Orig Health Dis* 9 (3): 266–69.

Thornburg, K. L., and Marshall, N. 2015. "The placenta is the center of the chronic disease universe." *Am J Obstet Gynecol* 213 (4 Suppl): S14–S20.

US Department of Health and Human Services. 2018. "Physical activity guidelines for Americans," 2nd ed. Washington, DC.

West, C. E., Jenmalm, M. C., Kozyrskyj, A. L., et al. 2016. "Probiotics for treatment and primary prevention of allergic diseases and asthma: Looking back and moving forward." *Expert Rev Clin Immunol* 12 (6): 625–39.

Winder, N. R., Krishnaveni, G. V., Veena, S. R., et al. 2011. "Mother's lifetime nutrition and the size, shape, and efficiency of the placenta." *Placenta* 32 (11): 806–10.

Xiang, Z., Yang, Y., Chang, C., et al. 2017. "The epigenetic mechanism for discordance of autoimmunity in monozygotic twins." *J Autoimmun* 83: 43–50.

Xu, M. Q., Sun, W. S., Liu, B. X., et al. 2009. "Prenatal malnutrition and adult schizophrenia: Further evidence from the 1959–1961 Chinese famine." *Schizophr Bull* 35 (3): 568–76.

Yan, J., Liu, L., Zhu, Y., et al. 2014. "The association between breastfeeding and childhood obesity: A meta-analysis." *BMC Public Health* 14: 1267.

Yoshioka, H., Iseki, K., and Fujita, K. 1983. "Development and differences of intestinal flora in the neonatal period in breast-fed and bottle-fed infants." *Pediatrics* 72 (3): 317–21.

CHAPTER 7

Allen, L., and Fountain, L. 2007. "Addressing sexuality and pregnancy in childbirth education classes." *J Perinat Educ* 16 (1): 32–36.

American College of Obstetricians and Gynecologists. 2009. Committee on Practice Bulletins — Obstetrics. "Practice bulletin no. 107: Induction of labor." *Obstet Gynecol* 114 (2, Pt 1): 386–97. doi: 10.1097/AOG.0b013e3181b48ef5.

———. 2011. Committee on Obstetric Practice. "Committee opinion no. 485: Prevention of early-onset group B streptococcal disease in newborns." *Obstet Gynecol* 117 (4): 1019–27. doi: 10.1097/AOG.0b013e318219229b. (Published correction appears in *Obstet Gynecol* 131 (2): 397, February 2018.)

———. 2018. Committee on Practice Bulletins — Obstetrics. "Practice bulletin no. 198: Prevention and management of obstetric lacerations at vaginal delivery." *Obstet Gynecol* 132 (3): e87–e102. doi: 10.1097/AOG.0000000000002841.

———. 2013. "Committee opinion no. 559: Cesarean delivery on maternal request." *Obstet Gynecol* 121 (4): 904–7. doi: 10.1097/01.AOG.0000428647.67925.d3.

———. 2016. Committee on Practice Bulletins — Obstetrics. "Practice bulletin no. 161: External cephalic version." *Obstet Gynecol* 127(2): e54–e61. doi: 10.1097/AOG.0000000000001312.

———. 2019. Committee on Practice — Obstetrics. "Practice bulletin no. 209: Obstetric analgesia and anesthesia." *Obstet Gynecol* 133 (3): e208–e225. doi: 10.1097/AOG.0000000000003132.

———. 2017. Committee on Obstetric Practice. "Committee opinion no. 684: Delayed umbilical cord clamping after birth." *Obstet Gynecol* 129 (1): e5–e10. doi: 10.1097/AOG.0000000000001860.

———. 2017. Committee on Obstetric Practice. "Committee opinion no. 697: Planned home birth." *Obstet Gynecol* 129 (4): e117–e122. doi: 10.1097/AOG.0000000000002024.

———. 2015. Committee on Obstetric Practice American Academy of Pediatrics—Committee on Fetus and Newborns. "Committee opinion no. 644: The Apgar score." *Obstet Gynecol* 126 (4): e52–d55. doi: 10.1097/AOG.0000000000001108.

———. 2019. "Committee opinion no. 766: Summary: Approaches to limit intervention during labor and birth." *Obstet Gynecol* 133 (2): e164–e173. doi: 10.1097/AOG.0000000000003081.

———. 2015. "Committee opinion no. 648: Umbilical cord blood banking." *Obstet Gynecol* 126 (6): e127–e129. doi: 10.1097/AOG.0000000000001212.

Berghella, V. 2020. "Cesarean delivery: Surgical technique." In C. J. Lockwood (Ed.), *UpToDate*. Waltham, MA: Wolters Kluwer Health Division of Wolters Kluwer. https://www.uptodate.com/contents/cesarean-delivery-surgical-technique.

Catling, C. J., Medley, N., Foureur, M., et al. 2015. "Group versus conventional antenatal care for women." *Cochrane Database Syst Rev* CD007622. doi: 10.1002/14651858.CD007622.pub3.

Declercq, E. R., Sakala, C., Corry, M. P., et al. 2014. "Major survey findings of listening to mothers (SM) III: Pregnancy and birth — Report of the third national U.S. survey of women's childbearing experiences." *J Perinat Educ* 23 (1): 9–16.

Edwards, M. L., and Jackson, A. D. 2017. "The historical development of obstetric anesthesia and its contributions to perinatology." *Am J Perinatol* 34 (3): 211–16.

Ehsanipoor, R. M., and Satin, A. J. 2020. "Normal and abnormal labor progression." In V. Berghella (Ed.), *UpToDate*. Waltham, MA: Wolters

Kluwer Health Division of Wolters Kluwer. https://www.uptodate.com/ contents/normal-and-abnormal-labor-progression?search=normal %20and%20abnormal%20labor%20progression&source=search_result& selectedTitle=1~150&usage_type=default&display_rank=1.

Funai, E. F., and Norwitz, E. R. 2020. "Management of normal labor and delivery." In C. J. Lockwood (Ed.), *UpToDate*. Waltham, MA: Wolters Kluwer Health Division of Wolters Kluwer. https://www.uptodate.com/ contents/management-of-normal-labor-and-delivery?search=pelvimetry &source=search_result&selectedTitle=3~11&usage_type=default&display _rank=3.

Grobman, W. 2020. "Techniques for ripening the unfavorable cervix prior to induction." In C. J. Lockwood (Ed.), *UpToDate*. Waltham, MA: Wolters Kluwer Health Division of Wolters Kluwer. https://www.uptodate .com/contents/techniques-for-ripening-the-unfavorable-cervix-prior-to -inductionsearch=techniques%20for%20cervical%20ripening%20the %20unfav%E2%80%A6search%20result%20selectedtitle%202%20150 %20usage%20type%20default%20display%20rank%202&source=search _result&selectedTitle=1~150&usage_type=default&display_rank=1.

Hodgins, S., Tielsch, J., Rankin, K., et al. 2016. "A new look at care in pregnancy: Simple, effective interventions for neglected populations." *PLoS One* 11 (8): e0160562.

Hodnett, E. D., Downe, S., Walsh, D., et al. 2010. "Alternative versus conventional institutional settings for birth." *Cochrane Database Syst Rev* CD000012. doi: 10.1002/14651858.CD000012.pub3.

Hofmeyr, G. J. 2019. "Overview of breech presentation." In C. J. Lockwood (Ed.), *UpToDate Inc*. Waltham, MA: Wolters Kluwer Health Division of Wolters Kluwer. https://www.uptodate.com/contents/overview-of-breech -presentation?search=overview%20of%20breech%20presentation&source =search_result&selectedTitle=1~150&usage_type=default&display_rank=1.

Hofmeyr, G. J., Kulier, R., and West, H. M. 2015. "External cephalic version for breech presentation at term." *Cochrane Database Syst Rev* CD000083. doi: 10.1002/14651858.CD000083.pub3.

Huseynov, A., Zollikofer, C. P., Coudyzer, W., et al. 2016. "Developmental evidence for obstetric adaptation of the human female pelvis." *Proc Natl Acad Sci USA* 113 (19): 5227–32.

Kennedy, H. P., Farrell, T., Paden, R., et al. 2009. "'I wasn't alone' — A study of group prenatal care in the military." *J Midwifery Womens Health* 54 (3): 176–83.

Leong, A. 2006. "Sexual dimorphism of the pelvic architecture: A struggling response to destructive and parsimonious forces by natural and mate selection." *McGill J Med* 9 (1): 61–66.

Long, E. C. 1963. "The placenta in lore and legend." *Bull Med Libr Assoc* 51 (2): 233–41.

Lothian, J. A. 2008. "Childbirth education at the crossroads." *J Perinat Educ* 17 (2): 45–49.

———. 2020. "Preparation for childbirth." In C. J. Lockwood (Ed.), *UpToDate* Waltham, MA: Wolters Kluwer Health Division of Wolters Kluwer. https://www.uptodate.com/contents/preparation-for-childbirth#references.

March of Dimes Foundation. 2013. "Birth plan." www.marchofdimes.com/birthplan.

Pattinson, R. C., Cuthbert, A., and Vannevel, V. 2017. "Pelvimetry for fetal cephalic presentations at or near term for deciding on mode of delivery." *Cochrane Database Syst Rev* 3: CD000161. doi: 10.1002/14651858.CD000161.pub2.

Rising, S. S., Kennedy, H. P., and Klima, C. S. 2004. "Redesigning prenatal care through Centering Pregnancy." *J Midwifery Womens Health* 49 (5): 398–404.

Rosenberg, K., and Trevathan, W. 2005. "Bipedalism and human birth: The obstetrical dilemma revisited." *Evol Anthropol* 4: 161–68.

Schultz, A. H. 1949. "Sex differences in the pelvises of primates." *Am J Phys Anthropol* 7 (3): 401–23.

Smith, R., Imtiaz, M., Banney, D., et al. 2015. "Why the heart is like an orchestra and the uterus is like a soccer crowd." *Am J Obstet Gynecol* 213 (2): 181–85.

Snowden, J. M., Tilden, E. L., Snyder, J., et al. 2015. "Planned out-of-hospital birth and birth outcomes." *N Engl J Med* 373 (27): 2642–53.

Thielen, K. 2012. "Exploring the group prenatal care model: A critical review of the literature." *J Perinat Educ* 21 (4): 209–18.

Vornhagen, J., Adams Waldorf, K. M., and Rajagopal, L. 2017. "Perinatal group B streptococcal infections: Virulence factors, immunity, and prevention strategies." *Trends Microbiol* 25 (11): 919–31.

Weiniger, C. F., Lyell, D. J., Tsen, L. C., et al. 2016. "Maternal outcomes of term breech presentation delivery: Impact of successful external cephalic version in a nationwide sample of delivery admissions in the United States." *BMC Pregnancy Childbirth* 16 (1): 150.

INDEX